Embarras Portage: Before, and After

Embarras Portage: Before, and After

Northern experiences of an aeradio wireless operator

Nicholas Verbisky
Edited by Amelia Breckenridge

Order this book online at www.trafford.com
or email orders@trafford.com

Most Trafford titles are also available at major online book retailers.

© Copyright 2005, 2010 Nicholas Verbisky.
All rights reserved. No part of this publication may be reproduced, stored in a retrieval system, or transmitted, in any form or by any means, electronic, mechanical, photocopying, recording, or otherwise, without the written prior permission of the author.

Printed in the United States of America.

ISBN: 978-1-4269-0748-7 (sc)

Our mission is to efficiently provide the world's finest, most comprehensive book publishing service, enabling every author to experience success. To find out how to publish your book, your way, and have it available worldwide, visit us online at www.trafford.com

Trafford rev. 08/12/2010

www.trafford.com

North America & international
toll-free: 1 888 232 4444 (USA & Canada)
phone: 250 383 6864 ♦ fax: 812 355 4082

Acknowledgement

With the completion of my writing there is a very special family friend I wish to thank very kindly for her assistance. Karen Compo has patiently spent many hours trying to decipher my hand writing, then to type it. For this I sincerely thank her.

Introduction

We all go through life's journey with a story to tell. For some time now I had entertained the idea of weaving the tale of some my experiences. They may not be of any great interest to everyone-however, it might arouse some interest among my immediate family and perhaps friends and acquaintances. My children in particular often have asked about the activities and events during my early years, not fully understanding certain milestones. There are events that are very vivid in my mind while others are somewhat vague. In any event it is my intension to provide some recollections for my children, grandchildren and subsequent generations.

Fifty years have now passed since some of my northern experiences occurred during the years 1952-1967. Being posted at Embarras Portage, and other various stations had a considerable impact on my early life. Many people will be named in the text; other will only be alluded to for obvious reasons. In all cases no libel is intended.

The lands and peoples one encountered during everyday life during that time, the joys and miseries of the northern communities, revealed the mysteries of a way of life which is rapidly vanishing. There were many remarkable tales to be told- of a way of living that existed nowhere else on earth, and in many cases that is now disappeared forever.

With a career change, my life's journey as a social worker took me through another twenty five years of service from 1967-1992. It presented me with a total change in lifestyle from my previous occupation. Helping and working with people less fortunate than most gave me a great deal of pride and satisfaction. There were periods of joy and on many occasions' disappointment, but there was also usually a balance that worked out properly in the end. The variety of the caseloads were so diversified that they lacked any degree of boredom. I am convinced I made the right decision with regards to my career change.

As I look back I see more clearly my legacy from my grandparents and parents. What a chance my grandparents had taken by choosing to come to their adopted land, and what a gift we received from them. Their toils and perseverance in a once desolate, inhospitable and unforgiving land is now a model for the modern world. They taught us how to survive during periods of difficulty, how to be strong and how to live a good life. Being brought up as a Catholic, my faith carried me throughout my life, whether things were good or bad. I consider myself lucky to have grown up in a family that taught me to be proud of who I am and what I do. I survived because of this legacy.

I thank God for my wonderful children; Anthony, Pamelia, Colleen and Colin as well as their spouses. I have nine grandchildren and I pray every day they choose the right road through life.

Here is my story.

Embarras Portage Posting

It was the first week of May 1952, and I was bound for Embarras Portage; my first northern posting. Very small chunks of ice still floated about in the Syne, which acted as a floatplane base. The Syne was located on the Clearwater River, a tributary of the Athabasca River. The Beaver Aircraft I was boarding was owned by Associated Airways, en route to their home base in Fort Smith N.W.T., with a request that I be picked up in Fort McMurray and off at Embarras. My drop was coordinated and arranged for by my new employer, the telecommunication branch of the Department of Transport.

The Beaver was loaded to the hilt, with the front seats occupied by Pat Carey, the pilot, his wife and their two small children in the rear jump seats. The family had flown out two weeks prior to Cooking Lake Alberta to change over the aircraft we were in from skies to floats and were carrying excess supplies. I was in the very back, crammed in amongst a bunch of boxes which the family was taking back to Fort Smith. Once the loading was completed, the pilot manoeuvred the airplane to the end of the Syne for take-off. Doing his flight checklist he revved the motor to the extent that it literally shook the whole aircraft. At full power we moved down the Syne with the airplane being rocked back and forward to produce the proper lift. As we gracefully moved off the water I glanced through the window- it seemed that we barely missed the tree tops. Once we reached cruising altitude of about two thousand feet the airplane leveled off, the motor returning to a dull monotone from a roar. I noticed the pilot talking on the radio being in contact with the Fort McMurray Aeradio, no doubt giving them his flight plan. During the flight to Embarras Portage my mind drifted back to the last week I had spent in Fort McMurray. It was my first view of a typical northern small town. I had flown in from Edmonton by Canadian

Pacific Airlines waiting for a pick-up, but my flight to Embarras was delayed as the airplane required additional repairs to the floats. I was told by our Edmonton Regional Office to check into a hotel and wait until another flight was made available.

The streets were wide, and the main street wasn't paved. My room at the Oil Sands Hotel was very primitive. It was about 8" x 8"- in it a small single bed, an ancient dresser and a wash stand with a basin and jug for water. One washroom down the hall served all the upstairs rooms. After I got settled in I wandered around town, having coffee at the café next door and hanging around the lobby of the hotel talking to other guests.

While in the lobby I struck up a conversation with a native fellow who seemed to be around the hotel a great deal. This fellow told me his name was John Shott. It was obvious he possessed a gregarious personality to match his massive size. He was huge, well over six feet tall and seemed to be in his late thirty's. He told me he was born and raised in Fort McMurray and worked on the northern transportation boats during the summer. John was familiar with the location of Embarras Portage, being the N.T. boats traveled by there. I asked him to join me for lunch at the next door café. At that time he gave me a brief history lesson on the Fort McMurray area. The town in itself appeared very poorly organized and very "northern" in nature. Travel by air or by N.A.R. rail was the only way in or out of this isolated place. A variety of small businesses lined the clay based main street and wooden sidewalks. Other then the hotel, there was a small drug store, a Hudson Bay store, a pool room, movie theatre, post office, and a two man RCMP Detachment along with a couple of cafes. There were few personally owned vehicles, but quite a number of taxi cabs for getting around. The only roads in the area were seven miles to the airport, bypassing the Waterways, a suburb of McMurray three miles away. The area between Fort McMurray and Waterways was known as the Prairie. It was the site of the Northern Transportation base along the Clearwater River. Waterways was also the terminal point for the Northern Alberta Railway. During my five day layover in McMurray I was able to get a fairly broad overview of the area, traveling occasionally by taxi.

Embarras Portage: Before, and After

On our flight we proceeded north toward Embarras Portage. I wondered what new experiences awaited me at this posting. I already knew some of what to expect. Once I had been notified that my transfer to Embarras was in effect, I made a point of obtaining records on the history of that location. The information I received told me that Embarras Airport was first built by the U.S. Army in 1942. Its main function was as an emergency and refueling station for the US Airplane movements within the north, and to provide weather reports. Wireless communication services were a primary function as well. It was maintained and operated by U.S. Army personnel until the end of World War II in 1945. Apparently, the large USA military presence in the north was due to a fear of Japan invading Alaska, which never did materialize. In 1945 this operation was turned over to the Canadian Army Signal Corp. They provided similar services, including landing service for the Forestry Department in the event of forest fires in the region. The airstrip at Embarras Portage was about 5000 feet in length, located on a level strip of sandy soil about half mile east from the Athabasca River. The Army Signal Corp operated the airport until 1951 at which time it was taken over by Transport Canada. There were two distinct branches involved in this case. The Civil Aviation Branch handled all duties pertaining to the operation of the Emergency Airport. Personnel at the airport included an Airport Manger, a mechanic and powerhouse operator, two equipment operators and a cook. The Telecommunication Branch, of which I was now a part, consisted of five Aeradio positions, one being the officer in charge. Embarras was classified as a compulsory reporting point for all aircraft movements in the vicinity or on Blue 84 airways. We provided around the clock communication to all aircrafts as well as recording and broadcasting weather reports. Weather reports were taken and sent by wireless radio to Edmonton each hour, mainly for the purpose of compiling weather forecasts for the pilots flying along the desolate northern routes.

Peering out the side window of the Beaver aircraft I noticed the river below. Mainly bush, muskeg, sandy hills and valleys covered this unforgiving landscape. Occasionally we would spot a moose munching in a grassy slew near the river. At one point the pilot banked and made a circuit as below he spotted an aircraft crash site; but on closer inspection, he could see that it was a small plane that had been in that condition for some time. Our scenic flight lasted almost an hour. Soon we were approaching Embarras, and the pilot made arrangements to descend. With the airport in sight, Pat contacted Aeradio for landing instructions. He was directed to land on the river adjacent to the airport property, which was on a high sandy bank. These instructions echoed back through the speaker, giving the pilot wind direction, speed and altimeter setting. *No reported local traffic and land at your own discretion.* The pilot acknowledged the report, crossed the airstrip and descended for a landing on the river with an approach against the current. The landing was smooth as he angled the plane toward a heavy wooden floating dock.

On arrival we were greeted by a number of airport staff. This was a big event. Not only was I a new face, I was arriving to fill an aeradio vacancy. This was also the first aircraft to land on the river since before the freeze up. During the winter the snow at the airstrip was rolled, making it suitable for the ski equipped plans to land. I was dropped off, and greetings were exchanged. The Beaver with the pilot and his family was soon moving down the river for take off to Fort Smith. A number of airport staff members happily helped me with my suitcase and kit bag up the steep embankment to the airport property. Charles Robinson was one of them. Charles was the officer in charge, perhaps in his 50's, and seemed like an easy going fellow. He had been in Embarras over a year and being short staffed was more than glad to have me on board. After a climb to the top of the bank I got my first look at my new home for a year or longer. The Embarras Portage airport was situated on an area of flat ground overlooking the river. As I carefully viewed the whole area, it seemed to me the main airport site wasn't that impressive. First, Charles assigned my quarters in a primitive quoin-set hut. He then gave me a general overview of the whole site. There was my place of work- a tar paper covered Aeradio shack. Two quoin-set buildings, one of which was our barracks, the other serving as a storage shed. There was also a power house, operating 24 hours per day, and a fairly large equipment garage, old and quite run down. The most impressive building that stood out was a large building operated and occupied by the Civil Aviation Department. This multi-purpose building housed their staff as well as serving as a mess hall, washroom and laundry facilities, and a lounge area for all the airport staff.

Aeradio Quarters and Duties

In comparison, our Aeradio operations buildings were not the most desirable to look at. The main room was only about twelve feet by twelve. There was a small cubby room about six by six which housed the officer in charge office and records room. The main entrance was through a small porch. In the main room there were two large transmitters. One was an AT3 which covered variable frequencies, and all voice type communications to aircraft. The other transmitter was low frequency used for wireless international code operation- best suited for C.W. communication and sending out the hourly weather reports by Morse code and all other messages. A console desk with a variety of radio receiving equipment, instruments and other controls was present as well. At one corner of the operations building were a conglomeration of equipment and charts required to create the hourly weather reports. Completing and recording weather was one of the major duties at this airport. A Stevenson Screen was located outside approximately fifty feet from our radio office. This covered and ventilated unit housed a number of dry and wet thermometers that gave current, as well as low and high temperatures for the day. From these we could also calculate the dew point. Another mandatory duty was to visually view the sky on an hourly basis to determine cloud conditions. By this we were able to establish the amounts as well as the cloud types. Weather reports are taken on the hour, recorded and sent to the Edmonton Control weather offices where I had been trained, for forecast purposes. Additionally these reports were sent to the adjacent stations of Fort McMurray and Fort Smith N.W.T. In return we also received their weather reports. All this was done by means of wireless radio using the Morse code. The three weather reports were than broadcasted simultaneously by radio phone and beacon on the same frequency at fifteen and forty five past each hour. Pilots normally monitored the radio beacons as they flew the Blue 84 airways, or if they were in the vicinity. The beacons were low frequency and each station on the airways had a different frequency allowing differentiation. During the winter this poorly insulated building was heated by a small oil heater. A great deal of heat was also generated by the transmitter and receiving equipment. This was reasonably comfortable during the winter months, but highly uncomfortable during the hot summer months.

Prior to the 1960's, all communications that required a quick exit within the northern regions was by means of wireless radio with the use of Morse code. Our department was required to provide these services to all government departments such as RCMP, Forestry and Northern Affairs.

After High School

After grade twelve at the Vilna High School in 1949, I had dreams of joining the R.C.M.P. I truly felt I would make a good policeman. There were very specific requirements for physical fitness and being brought up as a farmboy, physical fitness was the least of my worries. You had to be at least five foot eight, weigh 165, and the medical qualifications were also very high. After going through a medical with the local doctor I passed very easily. The next step which proved to be a barrier to my advancement was a written test followed by an interview at K Division about a month later. I felt I done fairly well on my written exam, but the interview was my downfall. It's fairly common knowledge that 50 years ago there was still a considerable amount of discrimination against any one that was not part of the English heritage. Of course, I did not fit into this category. My interview was conducted by a staunch British RCMP Officer, who obviously subscribed to this limited view as to what makes a good officer. Not being of the right nationality, I failed in this area and later received a rejection response. This was a very upsetting time in my life and still bothers me to some degree to this day. It wasn't the first or last time I had seen or felt the effects of racism in life. I didn't allow the disappointment to keep me down in any case. I was young and hell-bent for adventure. My parents, however, encouraged me to further my education. They claimed the farm life would be a poor choice. Many of my high school friends took up local training with CNR agent to qualify as telegraph operators which in turn would qualify them as station agents. I wasn't too interested in this area, but did inquiries about Radio Operator training at the Provincial Institution of Technology and Arts in Calgary. This was a 10 month course which offered students a second class radio license. The main training was in the area of wireless- requirements to pass were twenty five words per minute followed by extensive training in radio marine transmitting and receiving, as well as interdepartmental regulations pertaining to frequency worldwide. At the end of the ten months of training, we were required to pass very stringent Federal Government exams in addition to the college tests. Once a student earned a second class certificate, we would qualify for employment as radio officers on a variety of ocean-going vessels including the Merchant Navy. This would also make it easier to access career areas such as the Army signals Corps., or as a radio officer in the Air Force. After receiving information on this course from PITA, it seemed interesting enough and I applied. I was accepted and enrolled at the College for the September 1950 training.

Arriving In Calgary For Radio College

A few days prior to the start of college in Calgary, I boarded a bus for the city. I was leaving home for the first time with mixed emotions. A farm boy arriving in a big city was quite an experience. My parents, like most farm people living in the area, were very hard working but had little extra cash to spend beyond their basic living expenses. They were used to maintaining a hand to mouth type of existence. Fifty years ago there were no student loans, nor was the bank willing to loan us the money to pay for tuition or other student expenses. Upon arrival in Calgary I had the good fortune of staying with friends of my parents. They were very hospitable, putting me up for a week or so until I located housing closer to the college. At the college there was a bulletin board which posted room and board openings. There I located one at Third Avenue and Tenth Street northwest. The boarding house was run by an older couple who lived in a large, two story older home in a quiet neighbourhood near the Bow River. They took me on as well as four other boarders. My small quarters were in the basement. Crudely divided and finished rooms, with large furnace piping exposed along the wall, were assigned to each of us. Room and board rates were established at fifty dollars per month; this included breakfast and a bag lunch for school and supper. Apart from this, the kitchen was out of bounds. To be honest the meals were generally very sparse. As a boy I was used to hard work, but also used to eating hardy meals and there was no question of how often you could take snack or lunch breaks. Therefore the hunger cramp often existed. I couldn't just go out and buy the food I needed or wanted, or anything extra like that because I was given only fifteen dollars per month for spending, which was all my family could afford. I welcomed the occasional care package my mom would send to the house. My older brother, Stan, had started working in the oil fields near Redwater, and was generous enough to send me an occasional five or ten dollar bill- which was a godsend. As the months of limited food went on I lost some weight and I guess my stomach shrunk to the extent that the meals we had were sufficient to keep me going.

My residence was a mile and a half from the College. The last couple of blocks there was a steep hill to climb which we commuting students lovingly named Heartbreak Ridge. In our classes there were fifteen regular students from all over the province, and also eight veterans of World War II from all parts of Canada. The Department of Transport sponsored the veterans. The D.O.T. had a shortage of Aeradio operators. A condition of their sponsorship was they were required to spend at least two years working for the Federal Government. Like me, the regular students were all young and naive in many respects. The eight special students definitely had more life experience, and were outgoing and always seemed hell-bent for action of one type or another and willing to share that energy and experience with us newbies. Al Robertson was out main instructor. He too was a WWII veteran and extremely knowledgeable about the training. His expertise was as an International Wireless Operator, and regulations pertaining to international marine operations. Another instructor was Bert Bartch, an expert in the technology of the various radio equipment, both receiving and transmitting. In all areas of our training we were provided with very specific manuals that required equally detailed studies. I was always aware that my parents were working very hard to provide me with the necessary funds to take this training. I put all my efforts in to do well. This meant working diligently in classes as well as spending numerous hours at home doing home work. Carl, my room and class mate came from a rich family in southern Alberta. He was the only child. Spending money for him was never a problem. He drove a big BSW motor cycle and did not make any great effort in his studies. He was basically interested in receiving just enough knowledge to obtain a ham radio license. I would often hitch a ride with him to school and back. Another room and class mate was Alfred Sardina, who came from Trinidad in the northwest Indies. Alfred was thirty three years old and worked as a shoe salesman in his home country. He started training as a Commercial Radio Operator, but due to lack of some basic education he encountered a great deal of difficulty in many areas of training. After consultation with instructors at the college, it was suggested and he agreed to transfer to a less demanding course, Radio Technician Serving. Alfred was reasonably gifted with his hands, and did well in the new course. In the early fifty's there was a large demand for radio serving so he would have had no problem securing a job after graduation. Allan, another roommate was enrolled in an electrician course. He was a happy-go-lucky type of guy. He arrived there about a month after we had settled in. A very interesting event occurred with him at our usually quiet supper meal soon after his arrival. As I had mentioned earlier, our landlady was very careful in the amount of food that was put out on the kitchen table. She usually stood around and watched as we ate. On this night Allan was given a place at the kitchen table. The original residents were always very careful in assessing what was on the table and what each person should be allowed to take. Allan asked us to pass the food and loaded his plate up, leaving very little in the serving bowls for the rest of us. We nibbled on what was left, while Allan ate in a hearty fashion. The landlady was not impressed. After that at our meals we all were tuned into making sure we took our share, in a very prompt manner. It wasn't too long before Allan got the picture and settled into being considerate to the rest of us when it came time to serve up at dinner. Bud was our other roommate. He was an older fellow who was not a student, but worked for the City of Calgary. He had lived with this family for a couple of years and it seemed it was mainly due to the fact he received lower than usual room and board rates.

Attending college proved to be a hectic pace, but I enjoyed the training and the challenge. It seemed to be well worth all the effort. The instructors were very reasonable, making us fully aware that the end of this training involved federal government exams and they were providing us the essential information to pass those exams. Day to day our diverse class sometimes operated like a closely knit family, sharing a common interest of being successful in our studies. I did have periods of being extremely homesick. After about two months, a long weekend from school showed up and I was overjoyed to be heading home. Allan was going home as well. He had an old beater of a vehicle but it got him around. Four of us headed for Edmonton on a Friday afternoon with each one of us pitching in a few dollars for gas. Arriving in Edmonton I was picked up by my brother Stan and we headed for Vilna. It was wonderful to be with family again. I had a good time going out with a few of my high school buddies, taking in a dance at the local hall on Saturday night. Too soon it was back to the grind at the college.

Time went on and we all became more confident with what we were doing. Some had more difficulty than others. Based on my scores from the frequent exams we were given, I was clearly working at a much higher level than a lot of other students. I can attribute this to the many hours spent doing homework at the house and at school. Wireless Radio Code (International) was a demanding subject, mainly in receiving. Apart from class, living in Calgary was quite an experience for a small town farm boy. At first it was overwhelming, but as time went on I did get used to city life. The biggest hardship was lack of cash, and not being able to join in many outside activities although I was able to take in the odd college dance or go to a movie.

After a few months one of my classmates was hired part time as a bus boy at the Palaser hotel. He told me there were openings for more staff and asked if I would work there with him. It didn't take long to make a decision and I was soon on the job, mainly on Friday and Saturday nights. The Palaser was a very classy hotel, and work dress meant black slacks and bowtie, with a white shirt. Work included setting up and cleaning table, moving furniture, etc as well as providing car type of services to the cliental. The sixty cents per hour we receiving in wages was no great shake, but the fringe benefits were substantial. There were always many leftovers at the end of the night including gourmet type foods and we feasted on these at the end of the shift. Occasionally we would nip on some drinks which were always very plentiful.

The eight special students at the college were mostly a hard breed of fellows. The veterans all had some knowledge in commercial wireless radio, being each one of them had either served as with the Army Signal Corps. as a radio officer in RCAF, or in the Navy as a radio operator. The purpose of this additional training was that they obtained their second class commercial radio license, so they could be employed as Aeradio operators with the Department of Transportation Telecommunication Branch. These fellows did not study to any extent but with their back grounds and previous knowledge they really didn't have to. For the regular students our course was ten months, but the special students only attended for eight months. They wrote their exams at the Federal post office building. Six were successful on their first try. Two others required an additional month of training but later passed. Out of the fifteen regular students only ten of us successfully completed our training and written exams. The other five either dropped out or found employment with the oil field seismic crews as radio analyzers.

During the school year I had very little time for a social life due to the hectic pace of the student life coupled with a shortage of disposable income. I did meet a very sweet girl, whose name I do not recall. She frequently walked her dog near our home and I got to know her quite well. She was sixteen, attending high school nearby and was an only child. Her father was manager of a small cookie factory. We hung out occasionally, when I took some time off from studies. Usually on a Sunday I was invited to their home for dinner which was a treat, away from the wieners and beans that was often the staple meal at the boarding house. When I left Calgary I lost track of this young girl. After completing my exams successfully and knowing that I would be receiving my second class commercial radio license, I headed home in June. It was such a relief to get away from the stress of writing exams. Being home with family and friends was a joyful time for me, and I had missed them terribly.

During the later part of our training, our class was visited by a recruiter from the Department of Transport Telecommunication branch. He gave us a detailed summary of employment opportunities with their department. They had a severe shortage of qualified aeradio operators, mostly in remote and isolated emergency air bases throughout the north. During the school year a graduated program alumni would sometimes drop into class for a visit. They all were mostly employed by the Department of Transport, and usually on annual leave from some northern airport. Their experiences sounded exciting, with the good wages and fringe benefits such as free room and board it sounded like a good way to start my career. After spending a very comfortable and relaxing month with my family I applied for employment at the Department of Transport, in the Bowley Henry Building off Jasper Avenue in Edmonton.

Employment In Flight Services

After going through a short interview with the personal department, and being briefed on the conditions of employment, I signed up for service. Most of the bases were located in the north, a tough assignment for anyone, and a very detailed physical exam was required. I would also have to be provided with a number of months of on-the-job and classroom training before a northern assignment would take place. The commercial second class license I was going to be issued only qualified me for duties in the marine service. With the D.O.T. in addition to the wireless radio communicators I was required to undergo training in all aspects of air services. This included air navigation systems, meteorology, ground to air communication, response to emergency situations, air traffic regulations and procedures. To be an effective operator I had to understand communications from the ground, and from the sky. After signing up, Kevin Coyne, a senior Aeradio operator from D.O.T., picked me at the city headquarters and drove me to the Edmonton Municipal Airport Aeradio communications network station. It was located on the second floor of the airport terminal building. The sheer size of the massive set-up overwhelmed me. There were three operators on duty at one time sitting at consoles, and the various sounds that came through the radio waves appeared to make little sense. What I gathered as we toured the facility was that one operator handled all CW (wireless communications); receiving inbound weather reports from various northern stations and also sending reports out. Next to the C.W. operator was the radio/phone operator who handled all in and out bound aircraft transmission prior to or after instructions with the tower, which handled all communications with a five mile radius of the airport. Fifty years ago the air navigation systems were not what they are today; however, they were very effective given the technology of the time. A/C during that era moved a lot slower. The third or senior operator mainly coordinated the various operations, as well as filling in whenever the need arose. He was also assigned to provide training for new staff such as myself. There were always other new staff being assigned, like the eight special students who were well into their on the job training as they graduated a couple of months before me. Five classmates of mine from the college also signed up with the department. I was introduced to two other operators, Curt Hart and Herb Taylor and told to observe and ask any questions I wished to know. It must have been obvious by the expression my face that I was dumbfounded. All I could think was- how was I ever going to learn what was going on with all these operations? However, as the days and weeks progressed I got more familiar with the various operational functions and became more at ease. When there were enough students for instructors to proceed with meteorological training a start date was set. In addition to the aeradio students, regular meteorology students were also enrolled. The class consisted of lectures and the practical training of learning functions of the many instruments used to observe and record hourly weather reports. This training was six weeks. I found it to be very relaxing, as well as interesting. The trainers were quite friendly, a man and a woman. Both of them had very extensive backgrounds relating to meteorology. With the weather training now behind me, the next task was to learn how to decipher the codes that were being sent and received via C.W. and how to transmit via phone to aircraft requesting this information. Forecasts were also requested and sent in a coded form that required deciphering. Little by little these various forms of training were falling into place. Soon I was sitting in the C.W. operators position, and finally in the phone positions under the close supervision of our trainers.

Grande Prairie Assignment

After four months training in Edmonton, a decision was made to assign me to a moderately busy airport for additional in-service training. Grande Prairie was the destination and I was headed there soon after via the Canadian Pacific Airlines; which at that time were about the only major airline providing service to points north and along the northwest route as far as Whitehorse. The airport was located some 4 miles west of Grande Prairie. The administration building housed the various departments of the airport. There was the Aeradio operation office, which was very spacious and neatly set up. The meteorological observers office was next to it, than the airport managers office and a large waiting area and agents office for the CP airlines. The runway was paved with taxi strips and surrounding areas well groomed and maintained. Two operators handled the day shifts, with one assigned to each of the afternoon and evening shifts. A large part of the Aeradio officer duties were to provide direct service to the CP airlines. By C.W. we would receive aircraft cargo information dispatched in junction with the number of persons on boards, baggage information, fuel, etc. This would be forwarded to the agent who would then adjust flights leaving the airport. The people at the airport were very cooperative incorporating me into the system. I felt very secure at this new post, and was happy to be part of the friendly staff. In addition to maintaining a daily supervised shift, I was adjusting well to the off duty hours. The airbase included married quarters and a barracks for the single bunch. As time went on, I began to feel part of the organization. I was also being introduced to some of the characters of the airport. The cook and his helper worked in the single men's mess hall. They were always dressed in white uniforms. The meals were exceptional, when I compared it to the boarding room days in Calgary. At least I never had to watch how much was going on my plate! There were many things which opened my eyes to things that I had never been exposed to in the past. The cooks were binge drinkers, and every month or so were off on an excessive drinking binge for three or four days. As he lived in the single mans barracks the cooks severe habit was more apparent. When this occurred the cook's helper would take over all kitchen duties and provide liquor for his friend the cook. Once he recovered to his normal state, the helper indulged in some excessive drinking himself but never to that same degree. This process was very much known by the airport manager, but as they were such good cooks he did not fire them.

Two months after my arrival at Grande Prairie, I was starting to maintain a day shift on my own under close observation by a senior operator. Aircraft movements were monitored along the northwest station route on whose airways Grand Prairie was located. Weather reports were broadcast every hour on frequencies monitored by the aircraft flying with the region. All aircraft flying out of a specific area would file a flight plan. The departure would be noted as well as over flight and time of arrival specified. Most large aircraft departing would file with on IFR (Instrument Flight Rule) which required clearance and being monitored by the ATC (Air Traffic Control) which was most common if the weather conditions were not favourable. There were designated reporting points all along the routes. All aircraft had to report any delays or unusual occurrences to the operator who in turn would promptly advise the A.T.C. All departures and arrivals were also very promptly reported to AJC. Smaller type aircraft which did not possess the necessary instruments to fly IFR would fly via VFR (Visual Flight Rules). VFR being that at all times aircraft must have at least three miles forward visibility as well as one thousand above and below altitude. The Grande Prairie airport is located on the amber two airways, essentially a roadway in the sky. This is created by a network of radio ranger stations stretching in this case from Edmonton to the Alaska boundary with the last report in Canada point being SNAG. In all there was ten compulsory reporting points extending between these two points.

Aeradio Technology

Radio range stations consisted of a single non directional antenna that received continuous wave radio frequency energy from a transmitter, which would be voice moderated for weather or other information- or simply transmitted as an unpopulated carrier. The antenna, which was usually a single vertical tower, radiated this signal in every direction with equal strength. Four additional vertical antennas received a second signal from a second transmitter. The second signal was radiated 1,020 cycles above the carrier from the omni-directional antenna. The new signal was alternately switched between vertical antenna pairs in such a way as to produce alternate "figure eight" radiation patterns. The energy was switched between antenna pairs in Morse code sequence, to produce the letter "a" (dot dash) on one pair of antennas and the letter "N" (dash dot) on the other pair. By controlling the phase relationship of energy transmitted to each vertical antenna pair "legs" pointing in any direction are produced. Producing legs involved varying lengths of transmission lines as phasing networks. In the aircraft the pilot could select which signal he wanted. A three position switch marked "range-voice-both" was used. The very narrow filter would either notch out the 1,020 HZ range signal while allowing virtually all of the voice frequencies to reach the pilots ears, or the filter could be inserted in services to pass and prevent voice from being heard. In both positions, range and voice could be heard simultaneously. Phase difference between the received signals in the aircraft meant there was a narrow (one to three degrees) "on course" signal where both the International Morse code "A" and "N" signals were of equal strength. On the narrow beam, the pilot could hear both "A" and "N" and they both are combined to produce a sound like a steady tone, but if he moved to his right (either inbound or outbound) he'd hear one of those code signal against a steady tone of lesser strength. This area (which could be as many as nine degrees wide) was known as "the twilight zone". A twilight zone existed on both sides of the "on course" signal. A pilot flying within a twilight zone was always to the right of the course signal, just as autos travel to the right of white lines on the highways. The pilot is able to identify his exact position when he is directly above the range station thus being by passing over the "cone of silence" marker. Directly above the radio range antennas there was a vital total cancellation of all signals, because of the phase relationships of various signed components. This alerts the pilot and he reports the position to the Aeradio operator. The majority of aircraft flying the radio range airways are on an IFR flight plans. I had limited exposure to wireless radio because all airports on the Blue 84 airways had teletype systems installed, thus all communication including weather were sent and receiving by this means.

During my tenure at the airport I became friendly with one of the meteorological observers, Brian. He was a single chap, in his mid thirty's. He had served overseas in the air force during the war, as a fighter pilot. Flying was still in his blood and he was trying to get into flying with some commercial airlines. In any event, to keep up with his love of flying he managed to purchase a surplus air force trainer. It was a two sitter cornel which he got for less than $1000. It did require some upgrading, but this was carried out by a local air craft mechanic. Brian mentioned it was a little hard on gas, but flying the machine was well worth it. One day he talked me into accompanying him on local cross country flights. The aircraft had single front and rear seats, with controls from either position and was fully acrobatic. After taking it to a safe area away from the airport he put it through some fighter pilot manoeuvres. My body had some difficulty adjusting to this unusual experience, but I loved it and went up with him again and again.

After being in Grand Prairie for about five months I was considered a trained Aeradio operator, now suitable transfer to any stations within our region. Most single junior operators were stated for duties in the isolated northern areas. It was then that I received my assignment for transfer to the Embarrass Portage Airport. To say I was quite excited about this new adventure would be an understatement. I could only imagine what was forthcoming, the possibilities and unknown challenges ahead for me. First I had to go to Edmonton to undergo the very detailed medical and dental examination at the University of Alberta Hospital. After the exam, x-rays, and tests I was deemed suitable for northern isolated duty. The Department was very specific that employees must be in excellent health. Once you were stationed in a certain location; you must remain there for one year before coming out on annual vacation. I was glad to get home to Vilna to visit with my family for a week or so, before I got my orders. The department made the necessary travel arrangements. I would be leaving Edmonton via CPA (DC3) to Fort McMurray, than I was to make connection with a Beaver plane operated by Associated Airways enroute to Embarras Portage.

Life In Embarras

I was settling in for life in Embarrass with mixed feelings. First of all, there was no social life and being in my early twenties, the isolation alone was an adjustment. I figured that since I was going to be here for some time, I would try and make the best of it. Our staff were mostly an easy bunch of fellows to get along with. Around the clock operations meant we had to do shift work consisting of days, evenings and midnights. In the months and years that followed there were many staff changes. The original bunch included Chas OIC and Allan, who had been there a year or so; Joe who had been there five months; myself, the new boy on the block. There was also Ron who came in a month before me. Chas OIC was about forty five, a bachelor from eastern Canada. He had joined the department shortly after the war and was stationed in the Yukon before he came to Embarras. Allan was the senior operator had been with the department for a couple of years, with the past year being at Embarras. He was alumni from the College, I had met him on a visit once. Joe, in his late twenties was the student in our office. He came to Embarrass to earn some money and engage in some upgrading and this was a good place to study. Ron, the happy-go-lucky type, came from a radio family. His main interest was in maintaining ham radio contacts with all parts of the world. He spent most of his spare time doing just that. Our barracks, a hut originally built by the U.S. Army left something to be desired. There were six open stalls, two each located at each end and two in the middle. The two inch thick walls were poorly insulated and totally heated by a large oil heater which made it hotter than heck during the summer, and extremely cold during the winter. The summer months were also plagued with mosquitoes. Nets were a must over the beds. Apparently the department had plans to built newer, more suitable barracks for our staff but that plan was on hold until the next spring. The Civil Aviation Branch had more modern barracks in place, with a common dining room, laundry and shower facilities to which our staff, luckily, had access. Speaking of them, the Civil Staff were a reasonably good bunch of group to get along with for the most part; however they did display some resentment towards us which created a sore spot. They had been made aware that we enjoyed free room and board while their staff was required to pay a monthly fee. One thing I must say is on the whole we enjoyed excellent meals- but even this varied because there were frequent staff changes of cooks. A lot of these fellows suffered with alcoholism or other problems. When they first arrived, meals would be very good, but as time went by, they would begin to slip little by little to the extent they were either fired or leave on their own. The manager of the A/P Hugh was the only member who was married and the government did provide him with a nice three bedroom home a short distance from the A/P property. His wife and young son lived there, often travelling to the "outside".

Allans Boat

To keep ourselves from being driven to complete boredom, we spent spare time on our hobbies. One day Allan received permission to build a speed boat. He ordered all the necessary supplies from Edmonton, which arrived on the first barge delivery to the airbase. He was able to secure a small Quonset hut where he started building his dream boat. Allan also ordered a new 25HP Johnson motor. This size of motor was not known to exist in the Embarras area. All the trappers had big river skiffs powered by motors that were 8HP or less. In my off time I often helped Allan with the construction. It took shape after a lot of sawing forms and gluing and clamping. He did an excellent job, and after having it painted the boat turned out to be a masterpiece. It was in the river shortly after with the 25HP Johnson mounted for power. Usually everybody on the river traveled at a speed not exceeding five miles per hour. When Allan opened up his baby, it cruised well over 30 miles per hour. All the residents along the river would literally close their eyes when he went by thinking he was nuts to travel that speed, that he would surely kill himself. I often travelled with Allan in the front seat and the speed felt excessive even to me, especially on river travel. As we sped along at breakneck speed, there were times he had to dodge drift wood or other debris. It was thrilling.

The Lay of the Land

All along the Athabasca River there were registered trap lines, thus every few miles or so trappers with their families had their homes built on the river banks. Across the river was the Wood Buffalo Park. Trappers also resided there, but they had to be of first nation ancestry. All winter long and till early spring the trap lines were fully occupied but once late spring arrived all the trapping season closed. Some traveled to Fort McMurray or Fort Chipewyan to get away from the isolation and visit with friends, take a break from the lifestyle while other families stayed around and in many instances secured seasonal employment with the forestry department or working the river boats as deckhands. Some stayed at home, growing big gardens, catching and drying fish, or picking and canning wild berries. To escape the boredom Allan and I covered a lot of the river visiting families, who in most instances were very glad to have us for tea. They would always closely inspect our means of travel as no doubt it was their first exposure to a boat like Allan's. He had ordered from Fort McMurray four forty-five gallon drums of gas and a few cases of oil for the proper mix, which was plenty of fuel for us to cruise the river at leisure. As time went on we got to know many of the trapping families with whom we would spend a lot of our spare time learning about their life styles, their families and why they chose this isolated type of life.

Letourneau Family

The Letourneau family's trap line bordered the airport. Their home was located approximately a half a mile from the airport, walking along a wooded path along the high river bank. George was a Canadian born Frenchman who had come to the Embarras area forty years earlier. He first had trapped on the Lake Athabasca delta before registered trap lines were in place. After moving to this area, he was able to obtain his own trap line. George, now around seventy, was a soft spoken individual with a gentle disposition. As we got to know each other I learned that he was born in St. Albert, Alberta and spent his early days working at the Roman Catholic Mission as a teamster driving various missionaries around the region. When he got a little older he joined a survey crew doing work in remote areas near Fort McMurray. After that he decided to venture on his merry way, further north, and started his career as a trapper. From what I could tell he found his life quite fulfilling. George's Metis spouse "Grannie", and that the only name I ever knew her by, was also a warm and gentle person who always made me, as well as the other airport staff, feel very welcome in her home. There was always a pot of tea on the stove and she had many stories to tell about her life mostly spent in the Embarras area. The Letourneaus were George, Grannie, their eldest daughter Rose and her two children Gloria and Margeret, as well as Edna, George and Grannies youngest daughter. They were a happy bunch and even though there were some hardships, they all appeared to manage exceptionally well and there were never any complaints. The family had adapted to living off the land in many respects. They were always able to get an adequate meat supply either moose, deer or caribou. The river and creeks provided an abundance of fish which they either dried or canned. A large garden was also planted which provide a good variety of vegetables. In the way of fruit, wild raspberries and blueberries grew throughout the area in the sandy soil. Their log home, though small and sparsely furnished was adequate and comfortable for their needs. The Letourneau home was a nice place to visit and relax and just get away from sometimes hectic pace at the airbase. Of all the people we met, I seemed to spend the most time at their house. Having been brought up on a farm and been exposed to a harsh lifestyle, I found the way they lived was not much different from the way I was used to during my earlier years. I often helped out with their garden as well as fishing in a creek nearby. George's trap line extended some seven miles along the river and then approximately twenty miles inward creating a large rectangular area. During the closed trapping season, George often secured casual employment at the airport doing odd jobs, including being a kitchen helper. When fire season was on he would work for the forestry department. He would take off on the river by boat with a pack board and fire axe, covering a lot of remote areas which showed potential forest fire hazards as well as fires already burning. The pay was twenty five cents per hour, the days were long and he would often be gone weeks at a time. This would substantially increase the families yearly income, enough to carry them over the summer months when there was no money from the sale of furs. The family's assets were marginal. Other than their home they had a storage shed which housed most of the tools necessary for trapping. Leg hold traps of every size, from #1 for trapping small animals such as weasels and squirrels to #4 for trapping wolves. There was also #6 to trap bears. A variety of tarps and stretching boards, snares, trapping boxes, a fifteen foot river skiff with an 8 HP kicker, as well as two canoes, one regular size and another small, very unstable canoe. The canoes were canvas covered and every spring need a new coat of heated tar applied to cover any nicks and scrapes. There was also the dog team – five in total with a toboggan.

Major Forest Fire Outbreak

After being in Embarrass only a couple of months, a severe and very aggressive forest fire hit the general area. The landscape was dry and bleak and there was little rain at the time, which hurried the blazes progress. Firefighting procedures at the time were very primitive, and due to the many lakes and muskegs the area of the fire was difficult to reach. Within a week of being reported it had spread over a vast area. Smoke covered the airport, and hung in so thickly we couldn't see across the river. The wireless radio messages from our control centre were always busy, providing the forestry department at Fort Chipewyan on its progress. After the second week things did not improve. The smoke became so thick, that the McMurray air services Norseman which flew in mail and supplies to the Embarras Portage trading post could not stop by. Milt McDougal was a seasonal bush pilot who had flown in this area for many years. Usually he would stop at our floating peer on the return trip about twelve miles south of the trading post. It was always exciting to have him come in with some fresh produce, mail, tobacco and other incidentals we ordered from Hill's Drug store in McMurray. From our regular wireless radio communications with McMurray we had learned that the forest fire was also gaining on their airport and some evacuations had been ordered. Things were so bad that the government had requested the services of the Army to help fight these aggressive and dangerous fires. The heavy smoke created some respiratory problems for some of the staff. There was no fresh product left in the kitchen but we still had adequate supplies of canned goods. This forest fire was so intense that exploding timber literally jumped the river. To make matters more interesting, a major front of this fire was headed in our direction. The airport was closely surrounded by jack pines, and we knew too well that we were vulnerable. The forestry department sent some large pumps by river but without a supply of hoses there was little hope of saving much of the camp in the event the fire was upon us. Everybody at the airport was alerted to evacuate on very short notice as soon as it reached our door step. We all packed our gear and were prepared to leave. Our only access was by river. A large department crash boat as well as a few other private boats were tied up at the dock, ready to provide us safe departure in the event of an evacuation order. We waited patiently for the disaster to strike us, but it did not happen. A sudden major change in the weather saw the end of the disastrous flames. The rain came pouring down in buckets, and the wind shifted back blowing in the opposite direction into the burnt out area. The major fire was suddenly stopped six miles from the air base. The smoke still lingered in the sky as a grey reminder for weeks, but the air gradually cleared and we could breathe again. We thanked God for a return to our normal activities. While all this was happening one of the radio operators, Bob Poirier was due for a transfer. He was patiently waiting for the plane to take him to Edmonton via Fort McMurray. Bobs wife was waiting for him in Edmonton. Meanwhile, hot spots in the burnt out forest still needed to be permanently doused. George Letourneau and the other summer workers had quite a job keeping them in check. Only a small portion of Georges trap line was touched by the fire, with most of the ravaged and burned out land being just north of his line.

Aeradio Operator Resigns

After Bobs' departure, the station was again short staffed, which meant more overtime. Most of us welcomed the overtime and the bigger cheques that meant at the end of the month. It wasn't very long before Jim Taski was assigned to our station. He arrived on a regular scheduled flight out of McMurray. As usual Jim was greeted and welcomed to our group. He was young, a city boy and never been exposed to an isolated lifestyle. We did our best to be kind and help him adjust. After all, many of us had been in the same position when we arrived at Embarras. As kind as we were, there seems to be time for everyone coming here to be tested and initiated particularly working the mid-night shift. All alone in the station at night was at times a bit scary. On the midnight shift, every hour we were required to go outdoors about fifty yards to the weather recording Stevenson screen for data. The enveloping darkness and stillness of the northern night was often interrupted by the eerie cries of wild animals. Some of us were not above pulling tricks too, making scraping sounds on the aeradio office walls. This was the initiation ritual. Most of us handled it pretty well. Poor Jim could not adjust to this unforgiving and at times monotonous lifestyle, and returned to the city to pursue something "more civilized" as he put it. He resigned his post and could not wait for the regular scheduled run by MM air services. He charted a Cessna 140 to take him to Fort McMurray for $100 dollars.

Rabies Outbreak

It was a fact- Embarras was a lonely outpost. But there was always something happening to keep us on our toes. For example, one day a trapper from down the river came to our office to report that while he was in the woods he had been attacked by a fox. The offending animal was very skinny, shedding hair and frothing at the mouth. He had clubbed it to death. As all trappers were aware that an animal displaying that kind of behaviour could be suffering from an illness of some sort, he brought it in. When the airport manager saw this deceased animal, it was very obvious to him something was not right. By wireless radio we contacted authorities in Edmonton outlining what had happened. Very quickly we received a reply- their suspicion was the animal had rabies. We were requested to sever its' head, pack it into a metal container, seal it, and ship to Edmonton asap for proper analysis. Within a week we were informed that, indeed, the animal had rabies. We were told to use extreme caution in the area due to this outbreak. Soon other reports reached our office about animals acting strangely. The RCMP office in Fort Chipewyan was made aware of these problems and sent patrols along the river to notify trappers about the outbreak, and tell them to use whatever precautions were necessary. The patrol boat arrived at Embarras a few days later. The officers were glad for a break in the usual routine, enjoying the excellent meals and overall hospitality provided by the airport staff during their stay. We inquired to these officers about the possibility of purchasing side arms. When working the night shift there was always the likelihood of an attack by the prowling rabied animals. The RCMP officers agreed this would be arranged for whoever wished to purchase one. A few of the fellows got pistols of different calibers. I obtained a 22 caliber six-shot revolver. One of the reasons I got a small caliber sidearm was because the ammunition was cheaper and easier to obtain. I ordered a case lot, five hundred rounds, mainly to be used for target practice. I set up targets overlooking the river next to the aeradio office. I used up the first five hundred rounds quickly and this was followed by additional purchases. I became very proficient and accurate in the use of my new firearm. As the weeks went on, additional reports of animals with rabies were reported. One day a skunk wandered into the airport which appeared very disoriented. It was quickly destroyed but the smell lingered on for a long time after. George shot a fox which was skin and bone and half of its hair was missing.

Trappers Helper

By late summer no further incidents with rabies in the local animal population had occurred, and it was presumed that nature has taken its course and the disease outbreak was cleared up. On days off, I would usually spend most of my time with the Letourneau family. George had his trap line all mapped out with specific trails for his dog team. In the late summer he would follow these trails by foot, clearing out fallen trees and other debris. I took an axe and my six-shooter and followed him on this never ending journey into the northern summer wilderness. From May to July the mosquitoes are so aggressive, it was nearly impossible to go anywhere without being eaten alive. One thing our department was generous enough to provide us with was a green type of insect repellent; which did help, but was not fool-proof. Fall was approaching- then very few mosquitoes existed, but for the "noseeums" a type of sandfly to pick on us. After going with George for a couple of days the line cleaning was completed. A variety of hazards were cleared up. We had discovered some skeletal remains of animals which we presumed had died during the rabies outbreak.

Late summer meant berry season was in full swing. Wild raspberries grew abundantly along the airstrip, in the areas where dirt was pushed up against the bush. The sandy soil was also ideal for wild blueberries. They too grew in great numbers. As these berries were in season the whole Leterouneau family picked the fruits to be canned for winter. I was used to this type of natural harvest, having gone many times as a young lad, and often joined the family on berry picking trips. Our cook was always pleased when I would bring him some of these fresh berries- their aromatic smell and sun soaked flavor made for special pies and a break in the usual grub. One day while I was picking berries with another fellow from the base, we ran into a "little" problem. We'd found a spot where the berries hung down like grapes and the pickings were something to be desired. Evidently we weren't the only ones who thought so as we'd been joined by a small black bear cub. We heard some snorting going on nearby which we knew very well to be coming from its mother. Our departure was quick and furious, leaving behind our ice cream pails almost full of berries. The next day we returned to the spot to see if we could retrieve anything. We were not surprised to find the berries gone and our pails trampled.

Between shifts we often wandered outdoors from the radio shack, leaving the windows open and gains turned up on the equipment so we could quickly respond to any type of a request. Near the radio office overlooking the river below, we built a wooden bench. From there we could watch the large river boats pushing barges nearly every day. The boats, mostly owned by Northern Transportation Company supplied all northern and arctic residents with nearly every imaginable type of product. The boats out of Fort McMurray went as far as the "Willows", the end of the Athabasca River and the mouth of Lake Athabasca. The barges would be disengaged at the Willows and picked up by other boats that pushed them over the crossing portion of the lake near Fort Chipewyan, then down the Slave River all the way to the arctic. It was very peaceful to watch these boats go by, and they would always acknowledge their presence to us by blowing their fog horn. The boats full loaded going down stream with the four MPH current moved along at a good pace. On the other hand, the boats returning and bucking the current while pushing empty barges moved somewhat slower. Sometimes these returning boats would pull over to our pier to unload supplies. The captain would always welcome whoever was around to join them for coffee and doughnuts. It was a chance to visit with some new people other than the people we worked with day in and day out. These boats logged long hours from day break to sundown and would tie up for only a few hours during the darkness of the northern night.

Burdinski Family

Allan and his speed boat were kept busy running up and down the river as often as he had the free time to go. When I was off shift the same time as him, I would join him in exploring all the various channels and out of the way places. We made a point of visiting many of the trappers who lived along the river, who were pleased to have some company and always had the teapot on. The Burdinski family lived seven miles upstream from our home base. Bill and his wife Ann were in their fifties and had three children. George, the eldest, was in his early twenties. Ann was in her late teens and Rose being the youngest was about fifteen. Bill and his family had moved to the Embarras area twenty years earlier from a small town east of Edmonton. He had apparently got himself mixed up in some legal trouble and decided to get away from there. Bill was illiterate, so Ann conducted the business affairs for the family. None of their children attended school and Mrs. Burdinski did not make any effort to home school them. We later learned that the RCMP corporal at Fort Chipewyan did make some efforts to have them attend school, but the parents felt they were needed at home. The Burdinskis' home was located a half a mile up on a creek off the main river, a well constructed log cabin so neat and clean that you could eat off the wooden shiplap floor. In the yard there was a large storage shed for supplies and equipment necessary for trapping. There was also a huge garden located nearby which produced abundant vegetables. The soil was fertile, part of it was on an old river bed. There was an outhouse, as water and sewer services were non-existent, and a root cellar for additional storage. It seemed that this family was so well organized they almost totally lived off the land. They owned a large trap line which kept both Bill and his son George busy during the trapping season. Ann and the daughters were experts in preparing the pelts which needed skinning, stretching and drying. Each spring most of their winters catch would be bundled and shipped to a Winnipeg fur auction. They would keep some of the pelts to sell at the local trading post to purchase basic staple supplies. This is how they lived from year to year. They very rarely went to visit on the "outside", spending the summer months gardening and caring for their place. If and when there was an active forest fire within the area, both Bill and George would be requested to assist, earning them some extra cash. When Allan and I visited with this family I discovered they talked in Ukrainian amongst themselves. It was quite funny and interesting to hear them making certain comments about Allan and I. I was fluent in the Ukrainian languages, but did not acknowledge that at the time. It was only later the family learned of our shared bilingualism, to their chagrin. At the time of our visits Ann was being courted by a local trapper. Larry Daniels was building a log cabin across the river from the airport. He was native, and his trap line was located in the Wood Buffalo Park. The cabin site was on some high ground above the bank surrounded by a heavy spruce forest. We came to know Larry well, from his visits with us at the airport. Slowly the cabin took its form, creating a home out of the wilderness. The trees were cut down with a cross cut saw and skidded into place with his dog team. From time to time some of us would help Larry with the log lifting, but his dad Bill Daniels was his main support. Larry's parents and younger siblings lived about three miles down river from this site on the same side of the river and in the park as well. In a matter of a couple of months the cabin was complete. Though rustic in many respects, it was a solid and cozy home and soon after Larry moved in his young bride.

Embarras Portage Trading Post

Every once in a while trips by river were made to the Embarrass Portage Trading Post twelve miles downriver. We would pick up extras like tobacco, confectionary items, and mail that was not in the regular delivery. It also provided a break in the routine, and gave the delivery person a chance to shoot the breeze with someone different than usual. From conversations with the Trading Post workers, we learned the Embarras Trading post was established in the early 1920's, by the Reed family who had originally come from the U.S. The fur trade was established by that time, and with a large number of trappers in the area it was a reasonable area to build a post. With business flourishing, after a few years the Reeds decided to move to a new location where it was currently. Lumber had to be purchased and shipped in for the new store and living quarters which were built in the mid 1930's. At the same time the Reed's also built a post office which proved to be a real asset for the families living in the area. The Trading Post and post office were located on a sandy ridge that overlooked the channel of the Athabasca River. The Reed's operated the trading post efficiently for many years and from what we could gather treated the trappers respectfully and fairly. As income by locals was generated during trapping season, the Reeds had established charge accounts for their purchases off season. These would be paid off with the selling of furs. No doubt the price of goods at the store was considerably higher than on the "outside" but this was understood due to the high cost of freight, some of which had to arrive by air. As time went on Mr. Reed's health began to deteriorate until he was hospitalized and passed away in the early 1940's. Mrs. Reed required help in the store and hired a local trapper with a good educational background by the name of Gunnor Aspaas who had come to the Embarras area to trap in the mid 1930's. Mrs.

Embarras Portage: Before, and After

Reed with Gunnors assistance operated the trapping post and post office for many years, providing an essential service to all the local population. Gunnor passed away in the early sixty's and Mrs. Reed now well in her seventies decided to sell her business and return to be near her family in the US. A new family, the Peltons, than took over this operation which still stood on that sandy bank, keeping a watchful eye over the Athabasca river.

Emergency Medical Landing

Slowly the summer faded, replacing its warm northern breezes with the chill of autumn oncoming, and we were getting prepared for the long, cold winter ahead. The daily shift work carried on as usual, and with the arrival of fall some of the problems of the summer months began to subside. For example, in the summer sleeping during the day after a midnight shift was always a problem. Our Quonset hut barracks were very hot, and it didn't help having the nearby power house diesel motor noisily running 24-7. There were also the mosquitoes to contend with. With the arrival of fall, the bugs died down and the daytime heat subsided somewhat. Kind of got used to the power house noise but another problem was present. We still got our unwanted wake calls every once in a while, as the barracks was one continuous room and every time another person walked in during the day one would be woken up and there was no time of year that would not occur. Allan had his bed and quarters just next to mine. He was a hypochondriac by nature. His bed side table possessed almost every non- prescription medication you could think of. He often said he came north prepared to treat any form of illness that would affect him. Little did he know that in his zealousness he might cause a problem for himself and over time this self medication finally did catch up with him. The first sign was Allan became to complain about having some problems going to the bathroom. Various juices like prune juice were given to him. This did not help, and was followed by other medications which also did not help. After a number of days with no relief, his worsening condition caused anxiety and panic attacks for Allan. Further than that, his medical situation was beginning to alarm everyone around him. Charlie, our OIC, felt that with no relief in sight he had no choice but to wire message our regional director in Edmonton. He requested an emergency medical landing of a CPA DC3 which was on a regular flight southbound out of Fort Smith to Edmonton. This request for an emergency landing was passed on to the CPA management and granted for pick up later in the afternoon. Allan prepared himself for the departure and the rest of the airport staff were alerted for this special occasion. In the past there has been an occasional emergency landing by a CPA DC3 for pick up of an injured or severely ill Embarras resident. The one mile grass runway was adequate for landings by DC3 aircraft. After leaving Fort Smith the pilot radioed our station advising us of his ETA. Allan and the rest of the airport people gathered near the airstrip. I remained in the radio shack with Ron the radio operator on duty. About ten minutes out of Embarras the pilot indicated he was starting decent from 5000 VFR and requested landing instructions. Ron responded, *Roger checked your position. Wind NW 10 MPH. Alternates 2990 no reported local traffic, land at your own discretion.* The pilot responded, *Roger.* At this point I proceeded to the taxi strip to join Allan and the rest of the bunch. The DC3 made a low pass over the airstrip and circuit and was followed by an inbound approach. The landing was smooth and the pilot led the aircraft to the end of the taxi strip, and swung it back towards the runway. The starboard motor was momentarily shut down. The main door was opened. Allan waved goodbye and climbed aboard. Motor was restarted with a roar and the aircraft taxied away for take-off. The whole operation took only a matter of minutes with the aircraft again being airborne.

After arrival in Edmonton, Allan was admitted to the University of Alberta Hospital for treatment. He was back in Embarrass a week later coming via regular route, flying from Edmonton to Fort McMurray via CPA and then taking the regular McMurray air services scheduled flight. Allan did not comment too much on the treatment he received at U of A hospital, but sheepishly let us know he was fit as a fiddle with the instructions from his physician to try and refrain from too much of self medication.

Annual leave at harvest time

With fall at our doorstep and being with the department for one year, I was really looking forward to three weeks annual leave. I had no plans other than going home to see my family, and to help out with harvest operations on the farm. Getting home in time for the fall harvest and seeing all my family during leave was a happy time. Fall was always a busy time of the year on the farm, and a lot of it was spent getting ready to harvest and thrash the crops. All operations at this time involved a lot of hard physical work. First there was the cutting of the ripe grain by a binder, which than required stoking. The stooks provided for good drying which required two or more weeks of favorable weather, not raining or snowing. The final step was the threshing. I arrived home just in time for the threshing, which was good because it was the time when most extra help was needed. We had some mechanical help as well. As long as I can remember my dad always owned his own threshing machine. Our neighbours, who did not own a machine, helped us out and in return my dad would thresh their grain. Between four and six families, each operating their own rack and team, would assist with the threshing operation. It was hard but fun to work on the crew. Many of the fellows were around the same age as me, so it meant good companionship. We would get up at 6 and after a hearty breakfast get the horses harnessed up and fed. Thrashing began at 7, lunch at 10, dinner from 12 to 1, lunch again at 3 and quitting time at 7, followed by supper. All the hard work of thrashing required lots of sustenance. Dad would be thinking of moving the threshing machine from one family to the next depending on how dry the grains in stooks were, always making sure that each family had some threshing done in the event it would snow. After being up north for a year, I quickly found pitching bundles to be a tough physical job, and did require some toning of the muscles.

My job as an aeradio operator required little physical work, but I had always stayed active and worked out at the station. Still, it took three or four days to get into reasonable shape to operate a threshing crew. If and when the weather was dry, operations would carry on without any breaks; this could easily go on for three weeks or longer. When harvesting was in progress, everybody in the family from the youngest to the oldest pitched in to help. Mom and the ladies of the crews would always have the most difficult jobs. They would have to get up earlier than everyone to cook a large breakfast, followed by doing all the day to day chores such as milking cows, feeding pigs, etc. and on top of that prepare several large, well balanced meals for everyone on the threshing crews. I marvel to this day how my mom managed with all these difficult tasks, especially with younger children in the family that required her care and attention as well. My parents were always very giving to us; therefore I felt a sense of pride in being at home during this difficult time of the year to help them in any way I could. Dad would also hire local natives whenever possible to help out. With dry weather at hand, harvest was in full swing and this lasted a number of weeks. An occasional rainy day would dampen progress, but at the same time was a welcome rest for the crews and gave time for the realignment of the machines. During one of these breaks I took a couple of days off to visit some friends in Edmonton for a little socializing, taking in movies and a few dances. My annual leave was coming to an end and I again made plans to return to Embarras. Other then the inevitable homesickness, I was looking forward to returning to my old stomping grounds. I said my goodbyes and made my way back, following the same old schedule of regular flights, leaving Edmonton for Fort McMurray via CPA and catching the McMurray air services flight into Embarras. After a three week working holiday on the "outside" I was back at Embarras, happy to be with the gang and back to the old routine. The local trapping families were also glad to see me return. While on leave at one of the dances I had the chance to meet a charming young lady by the name of Nancy. We hit it off and she promised to write to me. To my surprise, she kept her word and I started receiving letters from her. Shortly after my return to Embarras I got the first one. I enjoyed getting news from the outside, and kept the communication going. She must have enjoyed it too, because we struck up a written relationship. Nancy got to know all about my life and the day to day activities which went on in a remote northern outpost.

Preparing for Winter at Embarras

In the late fall, the leaves on the trees were disappearing along with the plague of summer flies and mosquitoes. I enjoyed many hours spent walking along the peaceful country trails. I continued to visit frequently with the Letourneau family, who were getting ready for the upcoming trapping season. This was a very exciting time for all the local trappers. At the Letourneaus home, the garden produce was all picked and stored. Fish and meat caught during the summer was smoked and dried, and safely stored in a high cash to keep it away from the hungry local bears. Granny and the girls were busy canning wild blueberries, raspberries and saskatoons for the long winter ahead. George had put the canoes away for the winter, and already the creeks which flowed into the Athabasca were starting to form ice. The kicker boat remained out for the last few weeks before freeze up, in case a trip to the Trading Post needed to be made for some last minute supplies while river travel was still possible. In this respect we were lucky at the airport to have an emergency runway at our disposal. During freeze-up McMurray Air Services would change from floats to wheels and ski combinations, and in this way were able to maintain weekly flights in. The families who lived along the river would have to rely on their own preparations once the freeze up occurred. It was mid October and the snowflakes began to fall, accompanied by a strong north wind. The chilly air hailed to men and dogs the beginning of another trapping season. The local dog teams had been tied up since spring, and were ready for some action. George's dogs were extremely aggressive when he hooked them up for their first practice run. As excited as they were, it showed how out of shape they had become from laying around all summer. For the first few runs, they would tire quickly after a couple of brisk miles. To get their strength up, George needed to take them out on the trails regularly for exercise. I enjoyed helping him go on these practice runs. George started setting some of his traps in November. Squirrels were just one of the animals he specialized in. They were plentiful, and the income from their furs provided a large part of the families income. I offered to help trap the small animals; something I had done often as a youngster in Vilna. I would take out my Winchester 22 when we headed on the trails. It was a neat gun with a nine shot clip and mounted scope. My earlier target shooting paid off during the hunt, and I was pretty accurate in picking off whatever I shot at. George suggested to me that it was better not to damage the fur by shooting the animals in the head and eye area. Much to my surprise I was able to do that almost every time. The exercise, fresh air and activity were a great distraction from the often mundane pace at work. Sometimes my coworkers or Georges girls would join us out on the trails, hunting squirrels. Than we would bring a lunch, make a fire and drink plenty of fresh hot tea. These were very relaxing afternoons for me. Most days we would return with a pack sack full of these small animals. The work did not end with the shooting however. Many more hours would be spent by Granny Letourneau and the girls skinning and stretching the furs. It was my least favourite kind of work, so I mostly left it to George and his crew who were very proficient at it anyway.

Little by little, small chunks of ice began to form in the Athabasca River signalling the inevitable freeze-up. It was a sign to make a trip to the Trading Post and Post Office for the Letourneau family and all the other families along the river, although the Letourneau's were lucky since I could always arrange for the McMurray Air Services to bring anything they needed as they lived so close to the airport. Even so, George made a final trip in his large skiff, powered by a Johnson 8hp kicker. The trip took almost 2 hours going down with the current, and of course much longer on the return, bucking the current and with a full load of supplies. I enjoyed travelling with George in his skiff, on this much slower trip than in Allans speedboat. It was a different experience than cruising at breakneck speed in the summertime. The scenery was basically drab, with most of the leaves gone. As we went both of us kept our eyes peeled for wildlife. The other purpose of our skiff ride was to perhaps shoot and kill a moose, which George knew would be hanging around close to the water supply. Moose hunting season had just opened. We were a little disappointed not to see any moose on this trip; but George had his army 303 British rifle ready for action just in case.

Until the last possible moment river travel remained of vital importance, connecting the people who lived along it. Allan kept his speed boat busy visiting families like the Daniels, and I often joined him. Larry Daniels, who had married Ann Burdinski and built their log cabin home across from the airport was one of our most frequent stops. Ann kept the house as neat and tidy as her mother had. It contained only basic furniture, a woodstove, wood heater, hand crafted cupboards and a table which had seen better days. The bedroom was divided from the rest of the house with curtains. Much of the furniture could have used some TLC. In the middle of the woods like they were, however, who other than Allan and I would ever get to see it? Like all the trappers in the area, Larry was preparing for the season ahead. He had spent time getting his dogs back in shape for the trail runs. He had built new storage sheds and a high cash to protect food and furs from the hungry bears. The lumber used for these projects was usually salvaged during the spring and summer when the mountain run off would bring all kinds of debris, logs and many times loose lumber floating down the river which could then be retrieved for all kinds of construction projects. Apart from building the sheds, Larry had also managed to shoot a bull moose just a few weeks earlier. It was a custom around the area to share good fortunes in hunting and in any respect, and a small supply of meat was shared around. Eventually George managed to shoot a moose for his family as well, a supply the family really needed at the time. Georges kill took place during one of his frequent trips into the bush, near a creek down a fairly steep bank. It took the effort of his entire family to get the meat home. First there was skinning, which needed to be done where the animal was killed. The rest of the meat would be cut up in pieces small enough to haul home. Granny had sewn special pack sacks which could be mounted on the dogs, so they carried a fair amount as well. The moose carcass could not be left alone due to the number of predators in the area. A camp would need to be set up, with George spending the night to protect the meat supply. When I learned about this, Ron and I joined the family helping them with carrying the meat. We designed a stretcher and with some difficulty managed to carry a good chunk of it for them. After numerous trips and various ways we were finally finished packing the useable meat to its destination. Some would be given to the dogs for a treat, but most of it was preserved by either smoking or drying or being rubbed down with a salt like preservative. Regardless of the effort involved, the family was now happy and relieved to have an assured meat supply for months to come.

Winter arrived with a vengeance, a heavy snowfall heralding the cold weather and the North Wind blowing frequently making life somewhat uncomfortable. In this environment we stayed mostly close to home base. In close quarters, interaction between the staff was relatively good. One thing that made our captivity bearable was the excellent meals. That made it easy for all of us to add an extra pound or two if you weren't careful. As well as the regularly prepared meals, we also had full access to the kitchen for meals or snacks anytime. This was mostly to accommodate the shift workers, but we all took advantage of it. During the winter radio reception was not always well received, even with our high quality short wave receiver. The best time for incoming transmissions was in the evening hours. When we weren't working the fellows and I got together for cards and conversation, the occasional poker game. Other than that, many of us spent time writing letters to friends and loved ones on the "outside." It was always exciting when the mail plane arrived with letters that were addressed to you. Nancy, the nice young girl I had met during annual leave, continued to write frequently and I returned the correspondence. My family as well made a point of writing regularly to update me on their situation back home. With each new letter, after reading I would feel so homesick and wonder why I was spending my young years in a place such as this.

My Mother's Early Life

Upon returning to Embarras after three weeks with friends and family at the farm, I would often feel very homesick. When I felt that way my mind would sometimes drift off to the past, to some of the stories my mother had told me about her early life. The harshness and poverty of her life, as well as her family and neighbours lives, was difficult to comprehend. My mom was born in 1909 on her uncles' farm fifteen miles southwest of the village of Vilna. Her mom and dad worked in Edmonton but had purchased a homestead in 1908 four miles south. During the summer her dad, my grandfather, would walk to Edmonton- over one hundred miles- to seek employment on the railways. During any spare time he had he would journey back to the homestead to clear some land. While working for the railway he would earn ten cents per hour. Her mom, my grandmother, worked as a domestic in a hotel in the city. During this time mom lived with her grandparents in her uncle's home in the district of Downing. The family moved to their own farm in 1911, into a two room house that was built the same year. This house was constructed of logs and had a plank floor. It was a major improvement and a far cry from the families first home, which had been carved out of a hillside in a very primitive fashion. All the furniture in the house was handmade other than the woodstove. The bed mattress was gunny sacks filled with hay. Bread baking was done in a "peech" (outdoor oven) which was built out of a wood frame than covered inside and outside with clay. During the winter months when her dad was not working in Edmonton he would trap muskrats to earn some money for basic needs. Meat was available from hunting deer and moose in the remote, sparsely settled area. Fish was also very plentiful in the small lakes nearby. Their family would eat what could be caught. Most of the families in the area were of Ukrainian descent, but some non-Ukrainian families moved into the neighbourhoods as well. Everybody was poor and they all suffered the same hardships, all trying to make a living on new land with primitive tools. These rural back settlements allowed the new immigrants to keep their language, maintain their styles of food and clothing, celebrate their holidays and establish their own churches. This presence of Ukrainian neighbours was most appreciated by the women as they called on each other for assistance when their men left for work away from home. Non-Ukrainian neighbours were not in any way excluded when help was needed, but the language barrier was a problem. Ukrainian or non Ukranian, every spring the men in the district would gather, load on a horse drawn wagon and go off to work in the Edmonton area. The work would consist of pick and shovel labour building the numerous of railways in the area. They would not be heard of until they returned in the fall, no word on their progress or health, no word from home to them either. The women in the community were left to fend for themselves and care for their family and homes in every way. They were expected to plant gardens and look after their livestock. Grandpa had about four acres of land which was cleared in the first few years with a lot of heavy labour to do it. The trees were thick and plentiful; it took a lot of hard work to cut them down by hand. Then the roots had to be pulled and burnt. Wheat was hand planted, and once harvested had to be milled into flour by hand. Flour was a major requirement in keeping the family nourished. As time went on the family were able to acquire a cow for milk, a dozen of chickens and a couple of pigs kept to be slaughtered in the winter. Meadows contained a variety of wild grasses, which were cut by a hand sickle and stacked for use in the winter. Wheat was gathered in the same way. There were no stores in the area, so even very basic staples such as tea, coffee and sugar had to be bought either in Mundare or Vegreville over 60 miles away. Neighbors took orders from each family in the areas when making a trip to either of these two small towns. As a return trip with horse and wagon took four days, the men making these trips had to camp on the roadside each way. Money was always in short supply so most of the clothing items were hand sown, much of it from flour sacks. Wearing hand-me-downs was the common practice. Everybody walked bare foot during the warm season from spring to fall, or rawhide moccasins. Shoes were worn only for special occasions. Over time more homesteaders began to arrive in the Vilna area; a post office was established and finally a country store. It contained only very basic items, but residents were able to buy many of their needed supplies locally instead of traveling the 60 or so miles to either Mundare or Vegreville. Mom's parents and many other families around Vilna were very closely attached to their religion. They honoured the Greek Catholic faith, and the first Greek Catholic church was built in 1913 in the Stry district.

The first plots of land wrestled from the wilderness were small and farming them was very labour intensive. Oxen were the first draft animals to provide the motor power needed for plowing and harrowing. Planting of seeds was done by spreading it on cultivated land by hand. The ripened grain was cut with a sickle and tied into sheave. The thrashing of sheaves was done on a firm flat surface with a flail, the most ancient of tools used for this purpose. As more land became cultivated, better and more efficient farming practice were incorporated by the homesteaders. Grandad took up farming on a more permanent basis and by 1918 horses replaced oxen, and a seed drill and binder were purchased at a considerable expense. Whenever possible, the cost of purchasing equipment and machines was shared by neighbours. After some time spent daydreaming , my mind would snap back to reality and to the comparable luxury I enjoyed; even here, in the remote posting of Embarras Portage.

Trapping Season in Full Swing

All the trappers along the river were now fully engaged in tending to their lines for the season. Breaking trails by dog team and toboggan along the established routes was one of the most routine activities. Part of the expertise involved placing and setting trap sets in areas frequented by the many wild animals of the area. A favourite spot for many trap sets was along the frozen river and creek beds. Squirrels in particular were sought after, which for many of these trappers were a "cash crop", being plentiful and relatively easy to catch. As usual the catches were very good in the early part of the season and the hunters made the most of it, knowing the catches would slow down once the very cold weather set in and the animals wouldn't be moving around as much. The river was now frozen over so dog team travel over its frigid waters was a regular way to travel. Once on a good run the dogs could move along a decent clip, but the difficulty was mostly when passing another team. The usual way to handle this was one of the teams would move over to the side, with the dogs being held onto tightly until the other team passes. If this was not done, the dogs would likely end up in a fight with one and other. George's team had five dogs. They were all mixed breed, of the Heinz 57 variety. Dog trading between trappers was frequent, each one hoping to get the right mix to serve his needs with the most important being the lead dog. Due to regular trading, George's team was a little mismatched, but suitable enough for him. Although the trapping season was in full swing, some of the teams were still not in good enough physical shape to cover the trap lines every day. I was given the opportunity to work the dogs out when they were idle, and a trip for me to the trading post was planned. A well travelled trail was already established. It ran along the river, close to the shoreline where there was the least chance of thin ice in areas where the current was active.

Embarras Portage: Before, and After

On the day of my run, the weather although biting cold was clear and offered little wind. George's daughter Rose would be joining me on the run to purchase supplies and check the mail. She rode in the toboggan while I was in charge of "mushing". With a "mush" we started on our little journey. I hung on for dear life as we started out. White knuckling the handles, helpless to these dogs who were hell bent for speed. Inwardly I questioned my interest in agreeing to help. After the first few miles we slowed to a more reasonable pace, and their lack of fitness started to show as I had to egg them on at times to move a little faster. The entire trip to the Portage took a little more than 2 hours, including a few short breaks. When we arrived the dogs were tied up outside and given a small meal of dried moose meat. While they rested Rose and I headed inside to pick up the mail and visit with the store workers and other locals who had also dropped by. There were a number of letters for me to take back to the airport, and I also had a list of items requested by some of the fellows which I picked up for them. Extras like tobacco, papers and candy. Rose on the other hand was purchasing mainly staple items; tea, butter, sugar, lard, flour and for a treat a couple of rings of garlic sausage. As the Letourneau family treated me like one of their own, I didn't hesitate in purchasing them some confectionary items and a half pound tin of tobacco and roll your owns as well. Smoking was a common practice by most adults in the area, and a tin of tobacco was something everybody used to enjoy. It was one of the few creature comforts during the sometimes long hours of winter. The return trip from the trading post began much the same way as the trip there, with the dogs starting off at a breakneck speed and Rose and I holding on for dear life. Dogs are very smart animals, and certainly know when they are homebound. Even with the added weight of our supplies we started off at a good clip. Again, the animals slowed significantly after just a few miles. We took a couple of breaks to allow the dogs to rest, enjoyed a couple of hand rolled smokes with the new tobacco, and returned to the Letourneau residence with daylight still hanging in the sky. We hooked the dogs back to their lengthy chain in their individual houses. They looked exhausted after our trip. The supplies Rose had purchased were unloaded, and as always Granny had the tea pot on and we all sat down to enjoy a nice hot cup. It had been an exciting day for me, my first time mushing a real dog team along the frozen Athabasca river.

Oblate Fathers Visit

Once winter really settled in, the Oblate Fathers from the Roman Catholic Mission in Fort Chipewyan would start making rounds and visit the families who lived all along the river. The priest would leave his home at Fort Chip, travelling with a volunteer by dog team to the first known family where he would spend the night and hold services. The next morning, he would head to the next closest family and repeat the process, eventually arriving at the airport with the idea of heading to the Letourneau residence. Father Donto enjoyed his brief stopover at the airport, because of the excellent food and care he would receive at the base. Many of the families he visited were very poor, and unable to provide for him adequately but the priest enjoyed a hot bath, warm bed and other little luxuries they could provide.

Father Donto was a tall, slim man of French descent, with a beard and dark eyes. He commented to us on how the priests in charge at Fort Chipewyan restricted things like listening to the radio or owning a record player. He loved country western music, and was very pleased that we could offer him a chance to enjoy some of the records in our living area. We had a fun time, listening to western music and visiting late into the night. As the only practicing Catholic, I joined the father for services at the Letourneau home. It was wonderful to attend mass for a change and to share it with the Letourneau's, who had become an important part of my life. After a few days of relaxing and listening to country music, Father Donto decided to move on continuing his spiritual trek, visiting more families down the river. He headed down as far as Point Burley and Poplar Point, where about a dozen families lived. The turnaround point was 30 miles downriver. He would return in about 2 weeks for another brief rest before returning to the mission in Fort Chipewyan. For his trip the father always took at least 2 half pound tins of tobacco with him. One tin would be gone by the time he got to the airport, distributed to the grateful families he visited enroute. The other tin would be given out to the families he visited around Point Burley. Almost everyone in the tiny settlements he visited smoked. Two days into his two week trip, the tobacco would be gone. The fellows at the airport were very generous is providing Father Donto with smokes, which he very greatly missed by the time he arrived to us. When Father arrived back at our base on the way back to Fort Chipewyan, he was also in need of a good rest, a hot bath and to clean the few clothes he brought with him. After a few days of music and quiet relaxation, he again visited the Letourneau home for mass which I joined them for, and then it was back to the mission. George would provide a ride for him by dog team along the river path to the Daniels home. While he was at the airport I had plenty of time to talk with this man about his life as a priest. He told me that as a young boy in France he had always dreamed of becoming a sailor. However, it was the tradition in his family that someone enters the religious life. He felt that he got the calling, entered into the Oblate order and was quickly assigned to missionary placement in the northern part of Canada. He had been here in the Fort Chipewyan area for a couple of years now. Father Donto told me that he enjoyed the freedom of the northern lifestyle which was balanced out by the hardships one had to endure in this unforgiving territory. I felt that I understood exactly what he meant.

Hospitality Center.

Our airport was part of a so called "hospitality centre" in the area. Being located halfway between Fort McMurray and Fort Chipewyan, it was a well known stop over for the priests, government agencies, pilots and other travelers along the river. The summer was the busiest time for visitors. Small float planes flying either north or south found our base as an overnight stop. However there was no time of year that we couldn't expect an expected, or unexpected, guest. It was great to see the occasional new face spending the night with us. It was common knowledge there was good food at the airport, and a comfortable place to rest. One night a local pilot Max Ward, his wife and two children spent the night. Max made regular flights heading to Yellowknife where he was trying to start a bush plane operation. On this occasion he was flying in his four seated float plane. The Ward family stayed at the only married residence, Hugh Semple and his wife Lynns house. Hugh was also the airport manager of the Civil Aviation Department. Max's small bush plane company grew over the years into a major operation, starting with bush flying he later expanded into a charter operation which is now known as Ward Air. He was one of the characters who regularly took advantage of the airports hospitality services. A couple of other regular visitors were the forestry official and game warden out of Fort Chip, Jack Plews. We even had the law stop in on occasion, in the form of Corporal H. Ruthledge and his junior officer constable Jack Fargey. In the winter they would arrive by either dog team or bombardier- in the summer by their RCMP cabin cruiser. These fellows in particular were good company and had many interesting stories to talk about when it came to their northern service experiences. As the winter months passed slowly as the ice forming, we fell into a kind of a basic routine. I would often visit with the local families; Larry Daniels and

Nicholas Verbisky

his wife across the river in their log cabin, as well as his parents Billy Daniels who were another three miles downriver. The long nights were passed playing cards and listening to music on the radio. And of course, with none of these families possessing electrical power, it was coal oil lamps that we played by and battery operated radios we listened to.

Reports of Famine

January 1953 turned out to be a very depressing month. The extremely cold temperatures and continuous snowfall caused many hardships for the people in the general area. Trapping slowed down considerably due to the poor movement of the animals. Breaking trails and resetting traps was constant and difficult. About mid January, a native from Point Burley thirty miles down the river arrived at our station by dog team. He reported that ten families living there and in Poplar Point were without food and were in dire straights. Although things could be lean for some of these families in the best of times, the severe lack of food was being caused by very poor hunting and trapping due to the excessive amount of snow that had fallen. An urgent wireless radio message was dispatched by us to the Army signal station at Fort Chipewyan and forwarded to the RCMP. It was our understanding that the RCMP in Fort Chip had a good supply of buffalo meat which was harvested in the Wood Buffalo Park near Fort Smith N.W.T for situations like this. The native messenger, a soft spoken person in his forty's, was told to wait for a reply. His dogs were fed and he was given meals and lodging for the night. A reply was received later that night with notice that an RCMP Bombardier with an attached pop sleigh would be arriving the next day with the meat supply and other emergency rations as well. The next morning, after a hearty breakfast, the native messenger headed home to tell the families the good news. Corporal Harold Rutledge's bombardier arrived at our base just after noon the next day. The officer stated he had departed Fort Chip, which was only eighty miles downriver, very early but had experienced some delays. He'd had to buck a lot of snow and because of cold weather had to make some stops to warm up also. Harold was anxious to carry on with the trip. I was off shift on a day off, so I asked him if he could use some company. He told me he would be pleased to have me accompany him on this rescue mission. The bombardier and pop sleigh was packed with the frozen buffalo meat. He moved and relocated some of the items on the front seat and made room for me, which was a pretty tight fit. The bombardier is not the most pleasant thing to ride. Its engine is loud and exhaust fumes are something to be desired. The rough grind of the tracks and uneven snow with no springs felt like riding a bucking horse at times. It took us two hours to reach Point Barley. Now the evening was upon us and soon darkness would be contended with. During the trip and we talked about Harold's experience as an officer in Fort Chipewyan. I was fascinated by his stories, curious about the interesting life of an RCMP officer on northern duty. Harold was a seasonal RCMP officer who had spent twelve years in the north. He said he enjoyed serving in the north, for the last three years in Fort Chip. He was a tall man, in top physical condition and more than anything very dedicated to his work and was well liked by the people he served, most of them being of native decent. He was easy going and in no rush, which is quite common to people living in the north. He told me most patrols in the winter were carried out by husky dog team. The huskies were used by the RCMP through the northern regions. Harold mentioned being assigned for northern duty after completing his basic training in Regina. All young recruits would be sent to Fort Smith N.W.T. which was then the capital of the Northwest Territories. This was the headquarters for the G. Division. At any one time there would be about twenty young Mounties undergoing training for northern duties learning to handle dog team patrols and mastering canoes and river patrol boats for the summer. After six months the new Mounties would be assigned to various postings in the north which included stations in the far arctic. Harold said there was a fairly large island near Fort Chip located on Lake Athabasca known as Dog Island. During the summer the island was a resting place for the RCMP dog teams from various detachments in the north. A full time Special Constable was assigned to provide the necessary care and attention for these animals. As we rode, I wondered what it must have been like for that SC, to live the summer months with only the dog teams and the constant northern sun for company.

Soon we arrived at our destination. Point Burley was located on fairly level plateau which appeared in a high sand bank near the river. As we pulled up to the first cabin we noticed there were at least four or five other small log buildings in close proximity, to each of which a slab porch was attached. Stopping the bombardier, a large group of the residents approached us. The first to greet us was the native that sent the message to the RCMP. He appeared to be a spokesman and perhaps the leader in the small community. You could clearly see these people's gratefulness and relief to have the rations arriving, as many of them had been dealing with a non-existent food supply. Harold jumped up, and asked for directions for where to unload the supplies which was actually one of the log cabins we passed that served as storage shed. Harold then inquired about the residents at Poplar Point who had been encountering similar difficulties. He stated to the leader of Point Burley that they must share these rations with that group as well. It was a well known fact that natives living in the north in all cases share their food supplies if some families are not as fortunate as others. A number of the men and boys started unloading. We were asked to enter one of the cabins for tea, which was something we had looked forward to on the cold two hour trip. The porch of the cabin did not have a door but a large hide hanging over which had to be moved sideways to make the entrance. This log cabin had two large rooms, one served as the kitchen area and the second as the family's bedroom. There appeared to be an adequate supply of fire wood as the kitchen was warm. It was sparsely furnished inside with a simple wooden table, a well used wood stove which had a small cast iron flat top wood heater attached to the pipes. A few mismatched chairs and hand crafted cupboard which showed a few kitchen items. On the cast iron heater was a large water kettle and tea pot. On the floor sitting very close to the heater was a very elderly resident. A younger lady we assumed to be the wife of the native spokesperson welcomed for us to sit down. She spoke in poor English but stated we would be served tea. The elderly lady sitting on the floor spoke in Cree asking us showing a sign if we had any tobacco. As both Harold and I were smokers, we each had a good supply of roll-your-own tobacco on hand. I did not hesitate to hand over my tobacco pack to her. She promptly filled a pipe and began to leisurely smoke with a gentle smile and sign of a thank you to us. I gave this lady the balance of the tobacco to keep. It would seem that these people were not only without food but also did not have any tobacco. From what we were able to gather most of the other adults in this tiny village smoked and were without tobacco as well. The water kettle boiled vigorously and then a large handful of teas was thrown into the tea pot and briefly boiled. It was extremely strong but there was not sugar or milk to soften the taste. It still tasted very good, even served in not so clean cups. After this brief break to warm up one of the men came to state the supplies were all unloaded. In addition to the buffalo meat, two one pound bags of loose teas as well as a couple containers of instant potatoes and powder milk as also brought in. Harold made it clear that families in Poplar Point down the river were to share in these supplies as well. These other families were mostly relatives and Harold was assured that no difficulties would arise. We were able to get a count of residents at the village which was twenty three with slightly less at Poplar Point. Four elderly seniors and the rest of the children and adults would benefit from our visit. It was obvious these people were extremely thankful for getting these rations. Even before we left we were told large pots of meat were filled for boiling up a typical Indian stew. After spending over an hour we had made preparations for the trip back to the airport. We each rolled a few cigarettes for the road, than Harold handed over the balance of his pack to Charlie, the spokesperson of the village. He was thankful for the generous offer and indicated that it would also be shared with the smokers in his camp.

It was getting dark and we were hoping for no problems on the trip home. The temperature was only about 20 degrees Fahrenheit but the sky was clear and there was almost no wind. This meant our in-bound bombardier tracks were not blown away which would provide easier travel back go the base. Harold and I had good winter gear on and he had survival gear on hand in the event of some unforeseen trouble or breakdown. In any event there were trappers living on either side of the river between four to seven miles apart so we felt we could easily get to one of them in case of an emergency. We also knew there would be no dog teams on the river at this time of the night. Slowly we began our rock and roll trip home. It was so cold, and the clear sky was an icy black overhead. The department was generous when it came to providing winter gear, including an arctic parka, ski pants, wool shirt, fur hat, mitts with liners, sheepskin flight boots (military surplus type) and wool socks. I was wearing most of it, and from what I gathered the RCMP provided similar gear. The onboard heater pumped out enough heat to keep us reasonably warm. The moon began to rise slowly, still low in the sky. As it shone against the snow, the icy reflected light gave off an atmosphere of brightness. It gave me the surreal feeling of being totally in a world of our own, rambling homeward bound. The bombardier tracks made earlier in the day were easy to follow. Harold made sure we traveled near the shore line, avoiding any thin ice spots on the river which could be a danger. Occasionally we would see the dim light of a distant trapper's cabin along the way. I could almost smell the buffalo meat stew which must have been cooking back at Point Burley. We were hopeful the cook had some good leftovers when we arrived back at base. Two hours after leaving the native village, we made our final climb up the river bank to the airport. Gordon, the cook was still in the kitchen cleaning up after the evening meal. He was glad to see us and said he would have leftovers warmed up right away. We both ate a hearty meal, than sat down in the common room for some chit chat. Harold, with his broad experience in the north, had many interesting tales to talk about and we were happy for a change in the usual shop talk which goes on during the winter at the airport. Harold was assigned to one of the spare rooms in the civil aviation barracks, and we all hit the sack after a very long, cold day. Harold rose early for breakfast and made plans for departure back to Fort Chip. The temperature was in the neighbourhood of 25 F. He had some difficulties in starting up the bombardier which required warm up from a Herman Nelson heater. Harold sent a wireless message back to his base advising them of his approximate time of return, and zoomed noisily off into the frozen horizon.

Unrest At Christmas

Fall flew by and it wasn't long before the Christmas season was upon us. I expected this year to be a different kind of Christmas, my first without any family to celebrate with. Like many people I have always recalled the holidays as a very happy and peaceful time. My parents had instilled the importance of Christmas with its true religious meaning, the birth of Christ, as the major part of this very special occasion. Christmas dinner with all its specially prepared items remained clearly in my mind. Gifts from Santa were usually something we could wear. The lack of money was always a large problem any time of year, so toys were not something we received. Candies, nuts, oranges and apples were in plentiful supply for this one special part of the year. A Christmas tree was decorated on Christmas Eve with plenty of homemade ornaments, plus a few store bought ones Mom had purchased. We had no electricity in our home so no lights were added, but tinsel was used to brighten our little celebration.

My first year in Embarras, Christmas turned out to be a disaster and remains clearly in my mind still, over fifty years later. It all started peacefully enough. I was working the evening shift on Christmas Eve. All the staff were invited to the Leternoueau home for a holiday get together, including the civil aviation staff. I learned later that our fellows took over a few bottles of liquor which had been brought over for Christmas by the McMurray air service pilot. Our staff being only occasional drinkers, the Christmas season was one of the few instances when the boys enjoyed a spirit or two. In addition to this, Grannie had been busy making her special raspberry wine, which she had brewed earlier in the year special for the Christmas season. On the other hand Hugh and his staff from civil aviation were known to belt down the drinks more frequently. To make matters worse, the cook at that time went to the Letourneau residence with our fellows. It was shaping up to be quite a party. It was common knowledge on the base the civil staff and our members were not always on the best terms. A kind of rivalry and jealousy existed between the two groups, although living and working together, we usually managed to be civil to one and other. One of the main contentions had to do with money. As a perk for serving isolation duty we were provided with free room and board, while the civil staff was charged $75 per month. $75 at that time was a lot of money. Our monthly salaries were higher as well, and combined with the overtime hours, it proved to be a relatively high income. On the other hand, they also realized some of the extra hardships we encountered. They knew we worked shift work and our poor radio shack was not the most enviable environment. They also knew that we had undergone a lot of specialized training to be in our positions. Their staff, apart from the managers, were hired and learned on the job. So for the most part, civil aviation staff had a fairly balanced view of the perceived unfairness in our respective situations. I say "for the most part", because the distaste they felt toward us still reared its ugly head now and again. While Hugh and his boys were downing their drinks on Christmas Eve, our group was leisurely celebrating with the Letourneau family. I, on the other hand, was holding down the fort in the radio room until midnight. Afterwards I would be replaced by Allan, who wasn't much of a drinker. I was waiting for him so I could join the others at the Letourneau's and try to catch up on some of the holiday festivities. Hugh and his boys were whooping it up pretty good at the station by this time. They had discussed the fact that the aeradio staff and cook were at the Letourneau home half a mile away. Hugh got it in his head to hitch up his dog team and make a visit to the Letourneau's to see what was going on. He was two sheets to the wind, but managed to team the dogs to his destination. I was not there, but the story goes that Hugh called on George and became very belligerent with him although I don't know why. Hugh's behaviour did not sit well with some of our fellows and they got into a name calling incident. Hugh was known for being unreasonable when he was drinking. A shoving match ensued and finally punches were exchanged. Hugh was in excellent shape and could usually handle himself very well but because of his condition could barely stand up straight, let alone defend himself in a fight. Ron Studon was a skinny fellow and not very physically strong since all his spare time was spent on the radio with his contacts around the world. Despite his size he was also very sober that night, and managed to throw a few punches at Hugh to his face which eventually caused considerable swelling around the eyes. This hurt Hugh's pride even more than his profile. All this happened outside of the Letourneau home, with the dog team being in the middle of it. The dogs were disoriented by the action and became tangled in their harnesses. Some of the fellows decided to break up the brawl and helped to stabilize the unfortunate situation we were in. They were able to sort out the dogs and suggested that Hugh go home which he did, but he was extremely upset and promised that he and his group would retaliate and clean our clocks on Christmas day. This was not an idle threat- it could have easily happened. When my replacement arrived to take over the midnight shift I was completely dumbfounded by the events of which had gone on. Even though I was not directly involved with what had occurred earlier that evening, I knew I was still considered part of the crew that would answer to Hugh's humiliating defeat. The cook, who had made plans for a nice turkey Christmas dinner, refused to proceed after seeing firsthand what had taken place the night before. With the arrival of Christmas morning, we were hanging our heads very low, knowing too well Hugh and his boys would be coming after us and basically we were all afraid of the consequences. This Christmas day was even more depressing when I compared it to the many happy ones I had enjoyed with my family. Our OIC Charles felt this retaliation situation required the personal involvement of our headquarters. A fairly urgent wireless message was drafted and forwarded to the regional Director for immediate action to avoid any more violence. Even though it was Christmas day a reply was received in a matter of hours. This was addressed to Hugh, the Airport Manager firmly stating in a directive to refrain from any type of action that would cause harm to any of the airport staff, in this case our staff and particularly Ron who had been the one to embarrass Hugh in a fight.

The reply also indicated a D.O.T. inspector aircraft would be arriving a day after Boxing Day to sort out the problems that had occurred on Christmas Eve. So Christmas passed tensely for some but without any further fighting, and as planned the D.O.T inspection aircraft arrived the morning of December 27. On hand were four senior regional staff members, two from the telecommunication branch and the two from the civil aviation department. A meeting was held in the general mass hall with nearly everyone attending. After a great deal of discussion and ironing out some major difficulties, a better approach to getting along was reached by the supervisors to try and avoid more discomfort and bad feelings between departments. It was also decided to have Ron transfer to another station. Ron was given two hours to pack and he would go to Edmonton with the return flight. He was sad about leaving so quickly but knew it was in the best interest of everyone that he left. Ron was an avid ham radio operator who had spent a great deal of his off duty time talking to contacts around the world, so we knew we would likely hear from "VE6UB", his call name, again. The end result of the meeting were some new ground rules for all the staff. The D.O.T. aircraft, with Ron on board, departed later that afternoon. Even though many of the issues were supposedly ironed out, mistrust existed between our two departments as long as I remained in Embarras. Years later I was told that things did improve once new staff members replaced the ones that had endured this very unpleasant Christmas. It was a Christmas day I will always remember. We did eventually enjoy a turkey dinner when the cook felt safe enough to go back in the kitchen- even though it was a few days late. New Years day arrived and left that year with very few drinking related festivities.

Grandparent's Journey

The winter of 1953 was very cold and stormy. Other than working regular shifts we hung around close to the barracks. To escape boredom, some of us visited the Letourneau home to play cards and chit chat about the poor trapping conditions. When alone in my barracks at night, my mind would drift back to stories from childhood. I could only shake my head in wonder and admiration, thinking of the courageous pioneering spirit of my grandparents. My father's parents had told me some of the tales themselves when I was very young, how when they left Poland they packed only the most important items and sold the rest for steamship fares. How they sewed the money into their clothes for safekeeping, and scooped up the three youngest of their four children to an unknown land. They had gambled everything on a perilous journey to Canada in 1897. After a grueling two weeks on a stinking steamship, they arrived in Halifax. At least they had the company of people in similar circumstances during their journey, hundreds of other families determined to make a new life in this foreign land. From Halifax there was a train ride to Winnipeg which housed an immigration centre. Extreme hardship was an everyday occurrence. It was very overcrowded and of course, almost nobody spoke English. Like many of his fellow immigrants my grandfather was hired for building the railroad from Winnipeg to Edmonton. It was with the pick and shovel brigade, and meant long days with very little pay. As they moved west building the tracks, the men were gone for months on end from their families, only returning to provide them with some money for basic needs before returning to the work site.

While granddad was gone a very tragic thing happened. Two of their four children became very ill and with very limited medical assistance available they both passed away. A nine year old boy and twelve year old girl, my aunt and uncle. My grandmother had the grim task of burying them in Selkirk, Manitoba. Since there was no means of contacting my grandfather, he did not learn of their fate until months later. It was after this sad time the family decided to move on to Edmonton. They arrived in Edmonton and settled in Strathcona in 1899. There was only a ferry to cross the Saskatchewan River to the north side. Both sides of the river at the time were very sparsely settled other than some Indian camps on the river banks. Granddad was soon back working on the railroad, but his dream like most immigrants was to obtain a homestead so he could have some land of his own that would provide for a better life for his family. His income was very limited and he knew that in order to realize his dream he must save for tools and perhaps a team of horses and a wagon, as well as supplies to build a small house. In the meantime he explored some of the local areas he may want to settle in. In those days survey maps were neither complete or accurate, so a person needed to see the lay of the land for himself. Before they could purchase any land, however, my grandparent's first goal was to save enough money to pay for the fares of their other two children to come to Canada. Grandmother started working as a domestic for some elite families. Both the teenagers still in Poland, named Nick and Ann, were anxious to be reunited with their parents. It was not long before they arrived safely in their new land. I could only imagine the joy after they both were reunited with their parents. Another blessing came soon after when my dad Joseph was born in 1902.

In the years that followed, their daughter Ann became a young lady. She was courted by a young gentlemen who was a family acquaintance. This young fellow possessed a strong pioneer spirit as well, and had scouted out a homestead one hundred miles northeast of Edmonton, fifteen miles north of the North Saskatchewan River. It was in a district known as Stry. It was sparsely populated, but the people that had already settled there were mostly of Ukrainian/Polish ancestry. Initially trips were by raft down the river, then a wild walk along bush trails to the homestead property. Imagine the difficulty this journey entailed. Their daughter Ann got married and the family moved to their homestead. My grandparents were very happy for the young couple- to own land of their own was the most precious thing any parent could hope for.

In 1910 my grandfather Anton purchased his homestead in Stry for a $10 fee. To make matters interesting, his new homestead was just across the road from their daughters and son in laws quarter. During off season my grandfather made many difficult trips to his farm to build a small two room house. In the late fall of 1911 the family made the five day journey from Strathcona to their new home, and it was the realization of a hard fought dream. The move was by a fully loaded horse drawn wagon. On arrival they would be greeted by the unclear scrub which was now their home. They only knew a few souls and still couldn't speak a word of English. Like many of the hardy settlers who came to these parts they came with nothing but spirit and determination. They relied on hard work to build a better life for themselves and their children. I will be forever grateful to my grandparents for coming to Canada and giving me the opportunities I could only have dreamed of if they had stayed where they were. Anton was older than fifty and Mary close behind when they began their new life on the farm. They suffered extreme hardships scratching out a living with their hands in difficult conditions, and I can only imagine the loneliness and isolation they felt, the anxiety as they stepped down from the wagon. To realize their dream they gave away virtually everything they owned, and then made a journey away from their home country never to see their families again. Knowing my family I can only believe the one thing they did not do was complain.

On The Trap Line

As memories of my childhood replayed in my mind, pangs of loneliness brought on by isolation quickly brought me back to reality. The temperature at Embarras still hovered below the twenty degrees mark, but at least it had quit snowing, and there was a bright and crisp feeling in the air. In the afternoon one of the other fellows and I wandered over to the Letourneau cabin to visit. There was tea boiling on the stove, as always. George looked a bit like a trapped animal himself in that cabin. Trapping conditions were still poor but since it had stopped snowing George felt it was time to hit the trap line to reset traps after digging them out. I was off duty the following day and told him I would like to join him. I was back at the Letourneau home early the next morning. I dressed very warmly for what I knew would be a full day outdoors. The dogs, tired of being patient through the cold snap, were all jumping around and anxious to go. I brought along my 22, while George packing along his 30-30 Winchester. A tea billy, some sausage, tea and sugar were also packed by Granny into the toboggan. We knew we had to break trail on the established route. After hooking up the wild dogs we were set to get moving. I jumped into the toboggan and we were off in a flash, heading along the well marked but snow filled trail. George with his forty years experience as a guide maneuvered his dog team expertly, holding on to the rear of his sled with a "gi" and "ha" for the direction he wished for them to turn. Nearly all of Georges traps were set along a marked trail, mostly along the riverbeds where animals were most plentiful. About a mile from the point of departure we made our first stop. The location of the traps was marked by either a slash on a tree or a canvas ribbon. The dog team halted and we approached the set, but as suspected by George there was nothing in this fully covered over trap. The snow was cleared away, some bait was added and the snow near by was a brushed to give the appearance of nothing being present. Most animals are very clever, and would stay away from places when a human presence is noted. This setting was for a martin using a number one and half trap. As we moved along from one site to the next, it was mostly the same situation. We did hit pay dirt (or pay *snow*) eight settings later when we uncovered a well frozen lynx in a number 2 trap. George was very pleased with this lucky catch and it gave us some encouragement to move along further down the line. It was now well into the afternoon so we scouted out a place to build a fire and have some lunch. We found an area where there was large growth of spruce. George also had some snare settings for squirrels in this spot which had to be checked. The first item on our agenda though was to have some lunch. We came upon a previously used spot which he called a "hotel". It was sheltered by some fallen spruce used for sitting on, and there was a good supply of dry branches for a fire. The dogs were herded into a nearby spit for a rest. We soon had a fire hungrily consuming the dry spruce branches. It was comforting and familiar, like camping at one of the lakes near my old home town; to sit down, enjoy the crackling flames and light up a cigarette from one of the burning branches. We let the fire die down. While George was tending to his dogs, I got out the tea billy and filled it with clean snow, packing it down and getting it ready to put over the fire. With an axe I secured a green poplar pole through which I slid the wire of the tea billy. Constructing a crotch, I soon had the tea billy properly situated over the roaring fire. It was still very cold, despite the small camp we had set up. Hot tea would be a treat. I added more snow to the kettle and before long it was at a rolling boil. George threw in a large handful of tea and let it boil for a little longer before removing the kettle from the flames. We dug out Grannies bannock and sausage and warmed them over the coals. After resting nearly an hour we felt it time to move along the rest of George's trap line. It would be getting dark in a couple of hours. We made sure the fire was out and set out to scout the nearby spruce growth. There were a number of snares set in this area and we would check to see if there was any lucky catches. I grabbed my 22 in the event I might see something, but the squirrel movements in general were

Embarras Portage: Before, and After

not very visible, since not too many tracks were seen. George did find two squirrels in the snares which he immediately reset. Untying the dogs we were off again, entering and traveling along a very crooked creek. Sets in this area were mainly for mink and martin. Nearly all the traps here sat in the overhangs along the creek bank. One of the advantages to the sets location was that snow did not cover the traps unless very strong winds blew the snow into the crevasses, which was quite rare. After checking four sets, George was happy to find a large male mink, frozen but still intact. We were lucky for that, because often other animals would attack the trapped animals sometimes eating parts of them and damaging the fur. This was more common if the traps were not checked over long periods of time, when the snows are too heavy or the weather too harsh. It was dusk by the time we had finished checking all the sets along George's line. We were homeward bound. As usual the dogs knew they were in the final leg of their hard day's journey so they moved along at a good clip. I was still riding in the toboggan while George took on the rear of the sleigh. As we rode George described how trapping was very similar to gambling. For example, he said, today he had managed to trap a lynx, two squirrels and a mink, a reasonably good catch. On some trips he got absolutely nothing, while on other occasions he returned with a whole bunch. As we rode with the setting sun casting long shadows across his rugged and rough profile, the dogs laboured breath filling my ears, George told me he has been trapping in this unforgiving land over forty years; there has been periods of feast and famine, but he would not trade the lifestyle for anything in the whole world. This was a life which gave him total freedom. He was his own boss and did not have to answer to anyone. I could imagine how important this was to him, and how so many who chose to live around Embarras felt similarly passionate about the northern lifestyle. It was dark when we arrived at Georges home. We were greeted by Granny, Rose, Edna and Rose's two daughters. The girls were excited to see us back safely and Granny was glad with the catch. The animals were taken into the house to be thawed before they being skinned and stretched. I helped George get the dogs unhooked and each placed back at their own houses. All five dogs had small shelters were hooked by 15 foot chains from their collars to the entrance. At this point I headed back to camp just in time to eat a hearty supper. It was an exhausting day, although I enjoyed the exercise and fresh air. After supper I laid in my bunk and listened to some music on my short wave radio. At some point I drifted to a dreamless sleep.

Radio certificate, and Verbisky farm, Stry district aerial view

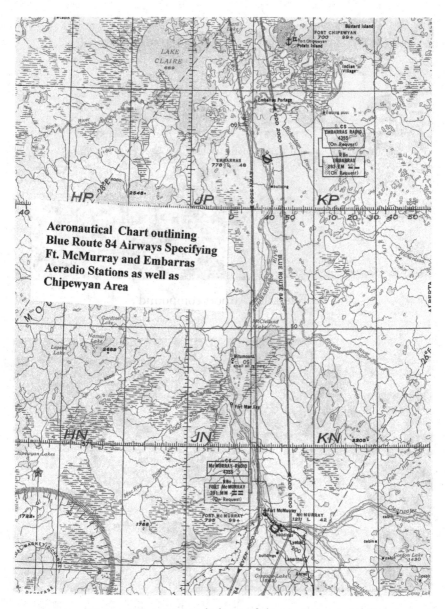

Aeronautical chart of the area

Embarras Airport compound

Students in commercial wireless radio training in calgary

Nick Verbisky on Georges trapline

N.T.C.L tugboat pushing barges on the Athabasca river

Charles Robertson, OIC Embarras Airport

Gloria, me, Edna, Margeret and George

The great beaver hunt

Rose and myself, dog sledding on the Athabasca River

R.C.M.P Corporal Harold Routledge stationed in Fort Chipewyan, on patrol stopover in Embarras Portage

Alan's speedboat

Fort Chipewyan

Downtown Ft. Chip

Uranium claim stakes

Safe arrival

Ron Studon, R.C.M.P special, Nick Verbisky
and Cst. J. Fargey, Embarras

Dinner break

Nick Verbisky, winter in Embarras Portage

Sonny Flett, myself by a trappers cabin near Fort Chipewyan

The Letourneaus and myself, in front of their home in Embarras Portage

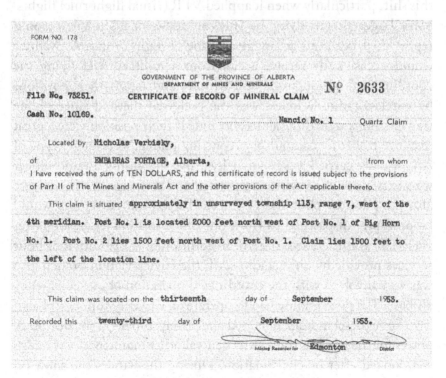

Mineral Claim certificate 1953

General Duties

When Ron left a few days after Christmas, we were short one radio staff and had to work overtime. Due to a high rotation of staff we were short staffed frequently. The extra overtime pay each month was nice and few of us complained. Work during the winter was for the most part routine. Occasionally an aircraft would report engine or mechanical trouble and request an emergency landing. These would be times when every effort was made to facilitate a favourable landing. Other than those rare occasions, day shift was the most productive time, as nearly all aircraft movement were on this shift particularly when it applied VFR (visual flight rule) flights. Some larger aircraft flying by IFR (intraconsent flight rules) would report their positions at any given time of night or day. As well we would occasionally receive a call from a military A/C flying the coordination's of longitude and latitude and report their positions by wireless radio. Wireless was the preferred method since phone transmissions were difficult to conclude if their position were a great distance from any compulsory reporting points. This longer type of A/C usually carried a crew of at least five, including a radio officer who was a qualified wireless operator and a navigator. There were also the pilot, co-pilot and engineer. These position reports would be promptly dispatched to from our office to military headquarter via the air traffic control system. In addition to the twenty four hour services provide by our aeradio staff, the Civic aviation branch also was very involved with the coordinated operation of our emergency air base. The power house had to operate on a twenty-four hour basis. Two large diesel motors, operated alternately, provided the necessary power for the whole airport. Any general other maintenance was also well looked after by civil aviation. During the winter, the snow on the runway was rolled down and packed. This provided adequate landings for both wheel and ski equipped aircraft. During the summer months, the runway was graded and the grass cut around the base of the air strip. A rotating beacon was mounted on a high metal structure which served as a very effective marker and could be seen for many miles. Kitchen staff were responsible for the control of food supplies and the like. So although work was routine, there was always duties to keep everyone busy even in the winter.

Aircraft Emergency

It was only February but the weather had warmed up considerably. One day after dinner I was in the barracks getting some of my clothing ready for a wash. Charlie, our O/C, was on duty filling in a vacant position. I would be going on the four to midnight shift. As I was moving about, I suddenly heard a low flying airplane which appeared to be approaching for a landing. This was not something that happened too frequently in Embarras, and I made a quick walk to our radio shack to see what was happening. Charlie was on the radio talking to the pilot who claimed to have an emergency. It was one of the McAvoy Bros., a bush pilot out of Yellowknife was flying his Cessna 195 to Edmonton with one passenger on board. What had happened was the front cable of one of his skis disengaged, causing the ski to flip into a vertical position. This created a severe problem in landing the aircraft. McAvoy had a large supply of fuel still on board and this could create an additional hazard. One thing that was going for this aircraft was the pilot was a seasoned bush flyer who had likely experienced similar incidents in the past. His passenger would need to count on that experience now. McAvoy advised us he would be making circuits in the vicinity of our station, lowering his fuel level to almost empty after which he would make an emergency landing on the snow rolled airstrip. This was potentially a dangerous situation, and all the airport staff was alerted immediately. This was not going to be easy for the pilot. If the landing was miscalculated even by a few degrees the vertical ski would hit the runway first, and at the point of impact it could cause the aircraft to spin and roll out of control. Catching fire was another real possibility that was known to happen in cases such as this. All able bodied staff at the airport including our aeradio staff had received some basic training in fire fighting and rescue. For basic firefighting equipment, a truck with a 500 gallon tank equipped with high pressure pump was on standby at all times. The station was also supplied with four wheel drive, surplus army jeep that was equipped with large cylinders of foam fire retardant. As the damaged aircraft made circuits in and around the airport, preparations were made to standby in the event McAvoy did not make a safe landing. Charlie continued to maintain constant radio contact with the pilot outlining what the ground crew had as a safety procedure in this emergency situation. Both fire control vehicles were parked near the ski airstrip with staff, including myself, ready and in place. After an hour or so of circuits he radioed in stating his fuel supply was depleted to almost empty and he would soon be making an attempt to land. He proceeded with two very low passes over the field to get his bearings. Everybody was on high alert as McAvoy was making his final approach. It didn't look like he would make a safe landing with the way the ski was hanging. Just before touchdown he cut his motor and banked in such a manner that he landed on the one serviceable ski. It seemed like an impossible task and at the very least could only be accomplished by an experienced pilot. His aircraft came to a sudden halt with no damage whatsoever. As McAvoy climbed out, a wide grin on his face, we all cheered his professionalism. His lady passenger seemed relieved that this ordeal was over and had a big smile as she exited the aircraft. After conversation with the airport manager the ski cable was properly re-attached. In no rush to take off again too soon, pilot McEvoy and his passenger were invited to the mess hall for lunch. The aircraft was checked over, refueled and in no time was ready to be south bound. Prior to departure he thanked everyone for helping them and standing by. We were just glad in this case that we never had to use our emergency training.

New Operator Assigned

One day we received a wireless directive from Regional headquarters that the long awaited aeradio staff replacement would be arriving at our base on the last week of February. As usual there was a bit of excitement that a new member would be coming. Mitch Stone arrived on the regular mail and supply run to Embarras Portage, on a stopover at our base with pilot Milt McDougal. Everybody that is not tied down to work greets the weekly aircraft to help unload mail and supplies. The pilot Milt was an easy going type northerner who always like to chit chat. Mitch was a tall, slim young fellow in his early twenty's. He was greeted by our OIC and staff members. My first impression of him was that his complexion was extremely fair and his hair was so blond that it was nearly white. After a lunch and meeting all airport personnel he was taken to his barracks. I don't think he was too impressed. As we continued the tour we could also tell he was not very pleased with the condition of our Aeradio operations office. Mitch went on to tell us he has been with the Department for two years now. The first year was spent in service training and then he was posted to Smith River B.C. where he had been for the past year. All of us pretty well knew where Smith River was, but were interested to know more details about that airbase. Many emergency airports stretched along the northwest staging route between Edmonton and Snag which almost bordered Alaska. It pretty well followed the Alaska highway by means of the Amber two Airways. Smith river and was only seven miles from southern Yukon boundary. Its gateway was approximately mile 533 on the Alaska highway. From there it was a 25 mile trip north along a difficult mountainous route. It was narrow with dozens of hairpin curves which required a great deal of care and attention. Mitch went on to tell us that the airbase was a very neat and well constructed site with mostly craftsman type log buildings. All in all, it sounded pretty nice. It must have been a great shock to see what the conditions were here in Embarras for Mitch. The base at Smith River was operated by the RCAF, as were most others along the staging route. The Aeradio air traffic central centre as well as the meteorological services were handled by our department. The RCAF staffs, both regulars and civilian, were in charge of all airport maintenance, 24 hour power operation, and messing (kitchen). He talked highly about the way this airport was operated and claimed very good co-operation existed between the RCAF and our department. The barracks offered large comfortable rooms and the tower in very good state of repair. Meals were provided by civil air force staffs who were considered to be excellent cooks. In the lounge area there was a RCAF Canteen which had a large variety of confectionary items as well as beer. Even though it was considered to be a fairly isolated base, it appeared Mitch's last post had been much more comfortable then what was offered here in Embarras. To arrive or depart Smith river required some team work and manoeuvring between the OIC's of Smith River and Watson Lake Yukon, the nearest airport that had regular commercial air service. Timing was set in such a way that one person would depart Watson Lake and also Smith River roughly the same time to meet at a specific location along the Alaska highway. There they would transfer a staff member, and depending on whether he was coming or going he would head one way to continue the journey. The main operations of both our central centers were very similar, except for one major issue. In Embarras all our point to point communications was by wireless radio. Along the staging route the landline communications was by means of a teletype system. Mitch had little C.W experience in the last year other than obtaining weather reports from an out port in Telegraph Creek. Mitch would be provided with a week or two of in service training before he would be given a shift. Mitch was a quick learner though, so it was only a few days before he was assigned to a shift of his own. Having a full staff reduced the overtime we were required to handle thus giving us normal days off once again for visiting and other adventures.

Hunting Squirrels

As we moved into the month of March 1953, the weather started to throw us all for a loop. The nights were bitterly cold, but during the day it warmed up a great deal if and when the sun was out. The Letourneau's thought this would be a perfect time to hunt squirrels which were on the move again, coming out after their winter hiatus. I had a couple of days off one week, and decided to go out with them for an afternoon of squirrel hunting. George, Edna and I planned to do just that on my first day off. Before we left, I informed the camp cook Gordie that I would be needing some items from the mess storage shelf for our camp. He was friendly with the family too and let me take a couple of cans of pork and beans, garlic sausage rings and some buns he had baked, for us to enjoy for a camp out lunch. These rations and my 22 Winchester I packed into my kit bag along with a good supply of shells. When I arrived on the day of the "big hunt", although it was still early in the morning, George and Edna were ready to go with the dog team lined up for takeoff. George would take the dog team to check the line, while Edna and I followed along on foot scouting the spruce growth for our tiny prey. The plan was to meet around noon at the "hotel" for lunch. The off trail snow was fairly deep still in some areas, but less so where the heavy spruce growths were, which would make it a little easier for us. We headed out almost immediately. It wasn't long before we saw signs of squirrel movement. I fingered my 22. George had given us very specific rules on the manners the squirrels could be shot. They must be shot through the mouth or eye area to protect the quality of the fur. If you missed and shot anywhere else, the pelt would be of little value when sold. My 22 was mounted with a very expensive scope, and with all the extra shooting practice I considered myself to be quite the marksman. Edna as well had practically been born with a gun in her hand, and with her 22 pump she did not misjudge many of her shots. As we moved deeper into the spruce grove, the high pitched squirrel chatter echoed through the trees. We positioned ourselves in a safe area and soon that sound was replaced by the popping of 22s. Within minutes, we had dropped over a dozen squirrels. We did not try to shoot all the squirrels that we could have at one time. It is the trappers way of securing seed for the future to not deplete stocks in any one location; maintaining the proper balance of nature. We packed up our bag and moved down the trail to another spruce grove. We repeated the same situation there, bagging over a dozen squirrels in a very short time. The third spot we entered was not quite as productive, and from there we moved to the "hotel" camp out location which was also in a large spruce grove. When we arrived George was nowhere to be seen, so we scouted the grove begging a few strays to be added to our collection. We cleared away some of the snow and started a fire, which blazed away as we sat down for a rest and couple of roll your own smokes. George arrived with the lunch supplies in his toboggan. He parked his exhausted dog team for a much needed rest. He, too, had been successful with his catches today. In the toboggan was a frozen coyote, a mink, and a half dozen frozen squirrels. He got up and pulled a frying pan out from behind a tree in the hotel. It was cleaned off with snow, and then we dumped the beans I had brought in there to warm up. We also dug out the garlic sausage and by now frozen buns and set them near to the fire. It wasn't long before we were eating, and did it ever taste good after all the fresh air and exercise. Afterwards we relaxed with a cup of strong tea from the billy which bubbled over the fire and another cigarette. After spending a good hour relaxing we decided to break camp and continue on out squirrel hunting. The squirrels Edna and I had shot earlier were added to the toboggan which made it a lot easier to travel on the snowy terrain. George headed out to finish checking the rest of his line, while Edna and I took a short cut back to their home. The return trip was somewhat slower, being there was still many areas of deep snow off the beaten trail. We did pass through a few locations of spruce growths where there were more squirrel communes. The population of these animals on Georges line were very good, as he had taken great care never to overhunt them from season to season. He had told me earlier that in a good year he would be able to get as many as 2000 squirrel pelts. Some would be sold to Embarras Trading Post for groceries and other supplies. The rest he shipped to a fur auction in Winnipeg. He usually received $1 per pelt which was a good chunk of his annual income. Edna and I stopped to hunt in two different areas along the short cut, and managed to add to an already productive day. Feeling pretty tired at this point, we decided we'd done enough for one day and continued to plow through the miles of snow homeward. When we finally arrived, Granny had the kettle on for tea. Soon after, George pulled in. The dogs looked as tired as we felt. After being detached from their harnesses, they just lay down to rest. After a tea break, the days catch was unloaded to the porch of the house. I knew some of the animals would be skinned later that evening, while others would have to wait for the following day.

While we were unloading, George mentioned that his frozen fish supply was getting low. He used this to supplement the dogs mash diet during the winter. He was hoping with mild weather to set a net in the next week. This would be near the mouth of a creek that was about a mile from the cabin. A strong eddy in the current there meant there was always a good variety of fish that would be caught. Even with warmer weather, however, setting the net was a tough job which required cutting through between 3 and 6 feet of ice. I could see how relieved George was when I offered to help him with the task on my next set of days off. After another cup of tea, I said goodbye and headed back to the airport. I arrived at the mess hall just before supper, feeling tired but satisfied after my unusual day of hunting squirrels. I have to admit, I did feel a little bad shooting all those cute little animals. On the other hand, I knew that their pelts were an important part of the Letourneau's livelihood as well as all the other trappers in the area. Since we were careful about how many we took, the balance of nature took care of itself and there were always more year after year. The day of hunting and trapping also was bringing back memories of winters from my childhood at home on the farm. During the winter I used to set many traps in one area where there were adjoining farms, checking my lines after school and

Embarras Portage: Before, and After

often not returning home until after dark. Weasel, jack rabbits, and bush rabbits were my intended prey. There weren't many squirrels around Vilna but during the spring, muskrats were plentiful. When the trapping season closed back then, I felt proud gathering up all my winter catches and taking them to the local fur buyer. It was interesting seeing how he would grade the furs and explain the prices that the different grades received. The fur buyer I dealt with in my youth had a good reputation and was honest in his prices, so I learned a lot from him during this process. The fur auction out of Winnipeg frequently sent out price lists as well so by comparison, we could see the local buyers prices were fair. Reflecting on this, with a full stomach after another satisfying meal in the mess, I decided there were in fact more similarities between my young life and this northern life that I would have previously thought before I decided to come here on this adventure.

Home Town History

Later that night, feeling a touch homesick, I was in my barracks resting. As usual my mind drifted off as I remembered some of the stories my parents and grandparents told me about the areas in which I was raised. Prior to 1918 the closest railway station to Vilna was Vegreville. The roads weren't really roads at all, more like rough bush trails with wagon ruts worn into virgin soil. The government had installed cable ferries in 1910, which gave the few settlers a reasonably convenient and safe river crossing in the summer. During the other seasons, river crossings were difficult and hazardous. From 1912 when the Canadian Northern and Grand Trunk Pacific Railways began to expand the government had promised homesteaders a railroad that would serve the area north of the Saskatchewan River from Edmonton to North Battleford. Plans were made and some surveying began, but WWI intervened and no tracks were laid on the proposed Edmonton to St. Paul branch. This branch line would have serviced mainly settlers north of the North Saskatchewan River. However, the rails had to run as far north as the present site of Vilna in order to avoid native lands, Saddle Lake Reserve and some sizeable bodies of water such as Cache Lake. The Canadian Northern Railway, later to become the Canadian Nations Railway (CNR) chose to avoid difficulties by passing between the Saddle Lake and Goodfish Lake Reserves. In fall of 1918 the railroad was built from the Oliver Junction in Edmonton to Vilna. The building of the rails were able to go as far as the terminal at Spedden thanks to the late arrival of winter. The following summer, the tracks were laid as far as St. Paul Des Métis where it stopped for some years. With the opening of the railway, early homesteaders could easily and efficiently shop their goods and order needed basic supplies. This meant avoiding the long and tiring three day journeys to the market places in Vegreville or

Embarras Portage: Before, and After

Mundare. These tracks become the life line of all the communities along its route. Meeting the train on its regular round from Edmonton a couple times a week was a most exciting experience and served as an important community gathering place.

Net Fishing

In March of 1953 the weather was holding out fairly well, with mostly cold nights balanced out by the comfortable, sunny days. A couple of weeks after squirrel hunting, George felt the weather was mild enough to set nets to gather a much needed supply of fresh fish both for the family and the dog team. Gung-ho to experience something different, I was happy to tag along and help. Netting fish was something I had never done. I was on two days off before going on night shift. George warned me ahead of time we had some pretty hard physical labour ahead of us, chopping two large holes in the thick, early spring ice. Luckily for us George knew where the ice was thinner over an eddy where the river entered the creek, and that was where we would drop the net. Right after breakfast at the mess I walked over to Georges home. He was already packing items for the trip into the toboggan. There was a fifty foot net, a couple of axes, a jig for running the net between the two holes in the ice, a sharp metal bar, a shovel and some other things. When all items were packed, we went into the house and kicked down a cup of coffee before hooking up the dogs for a short one mile trip to the intended spot. When we arrived George marked out the exact spot. Since he had used this location in previous years so he felt confident it would be productive. Then we started chopping away at the ice. The holes had to be large enough to let the jig enter under the ice. If I recall they had to be at least three feet square. George had been serious in stating that the chopping was going to be hard work. The ice was at least three feet deep. First we used an axe and as we got deeper down we used the metal bar which had sort of an ice chipper attached to it. After a hour both taking turns we managed to break through the ice and the hole being quickly flooded with water which was a good three feet deep. We took frequent rest breaks, and during one of them he told me how lucky he was for me to be helping him out with this difficult but necessary chore. I could see why he felt that way. The jig is painted red so it could be seen through the ice while it moves from one hole to the other. It is quite an act the way it is fed between the two holes, with the net being attached while it moves through. The lining up of the jig to travel between the two holes requires patience and persistence. After three hours of a lot of hard work the net was set in place between the two holes. Markers were set up, and we were done. The net is normally checked once a day, but George was excited about his new set and said he would like to check it just for curiosity later that evening. After lunch at the cabin I headed to the barracks for a rest as I was pretty well exhausted after all the chopping that took place. I told George I would be back in the later in the afternoon to check the net. George, Edna and I returned to the net site late in the afternoon. The pulling of the net also requires a great deal of know-how and patience. Ropes are attached and the net is pulled out slowly. To my surprise, George was in luck with a small catch. This included two ling-cods and a large northern pike. The net was reset. Things were good. The family would enjoy the fresh pike for supper, the dogs would feast on the cods minus the livers. Cod livers are considered too important to be given to the dogs instead served as a health food, mainly for children. The nets would be checked daily and remain set until a suitable supply of fish is caught to serve the needs of the family and their dog team. In the days that followed I joined George with the net lifting whenever I was available. Two other of our staff members Mitch and Al had never seen fish netted in this fashion, and also tagged along. There was always a variety of kinds of fish in the nets including whitefish, pike, cod and pickerel. The types and amounts varied from day to day. George, Edna and I would play a guessing game about what type of fish we would find each time, and how many. George was very generous in giving us fish for our mess hall. It was nice to have something different for dinner but Gordie, the cook, was not very anxious to do the cleaning so we had to take on the chore. It was worth it. After only a week and a half, George felt that he had an adequate supply on hand to last until breakup of the river which would be somewhere about the first two weeks of May. To keep the fish fresh he had a large box constructed with alternating layers of fish and snow packed in, a primitive northern refrigerator. He would be setting nets after breakup in the same location, with the process being a lot simpler in open water and the catch not requiring cold storage.

The weather at the beginning of April was unusually kind. The sun was moving up in the sky during the day bringing warmer temperature and causing some snow melting. It was a real pleasure to be outdoors during this month. On any days off work, I continued to spend a lot of time at the Letourneau's house. We would take off into the woods hunting squirrels and checking the lines with George. With the sunny and warm weather the animals were on the move and so catches were good. George and Granny were two very busy people when it came to skinning and stretching pelts, doing this by lamplight every evening. We all started to notice more dog team travel along the river. The trappers from up river would be making trips to the Trading Post at Embarrass to sell furs, buy supplies and pick up the mail. Sometimes a lone trapper would stop at our station for a chat, rest his dogs and have lunch with us. Others would stop to send a wireless message to the outside. A few of the natives from Point Burley and Poplar Point where we had done the earlier delivery of the buffalo meat also stopped by to let us know they were once again making good progress with their trapping and hunting, and did not require any further assistance. They asked us to notify Corporal Rutledge of Fort Chipewyan.

During the nice seasonal weather everyone seemed to get out and move about more, helping many of the guys get over the long winter blues. There were days that George was swamped with skinning and stretching pelts that the dog team was free. Edna, Rose and her two daughters, would want to visit with their uncle Bill Daniels and his family who lived in the wood Buffalo Park three miles across and down river. Al and I would join them sometimes, but we would have to walk there. The dog team trail was usually well packed so walking there was good exercise. The whole bunch of us would get together for a nice visit, play cards and end up with a lunch of tea and bannock. Sometimes we would visit with Billy's son Larry Daniels and his wife

Embarras Portage: Before, and After

Ann, who lived across the river from the airport in their own little cabin. Visiting with the local people when we weren't working made life in Embarras Portage a lot more colourful.

Medical Evacuation

It was sometime around the middle of April 1953, when a trapper from Polar Point arrived at the radio office. This young man was accidentally shot through the shoulder by a 30-30 rifle. When he arrived at our airport, he was wrapped up in a toboggan and told us he was in a lot of pain. He was with a native fellow in his late twenties. We could tell by the excruciated look on both their faces that it was serious. The airport manager was trained in advanced first aid, and ordered us to take the injured man to his office for an examination and emergency treatment. He had lost a lot of blood and the injury was severe enough that he required immediate hospitalization. Al was on shift and dispatched an urgent wireless message to the RCMP in Fort McMurray requesting a plane to pick him up. In the meantime, the injured arm was well wrapped and a sedative was given to him to ease the pain, which it did because the trapper relaxed and nearly fell asleep. The McMurray Air Service ski plane piloted by Milt McDougal arrived within a couple of hours. He was joined for the flight by a nurse and CPL Dagg of the RCMP. The young trapper was carefully loaded into the aircraft and they were off in a flash. CPL Dagg received a full report en route of the accident that had happened from the fellow that brought him to our station.

My Early School Days

Alone in the barracks, when I wasn't visiting the Letourneau family or writing letters to my family and friends, my mind would sometimes snap back to my childhood years. Growing up on a rural farm in Vilna had given me many vivid and wonderful memories. I clearly remembered the days of walking one and half miles to Stry School over snow drifts that completely covered the fence posts. Uphill both ways, I actually lived this old cliché. Horse travel was difficult in the winter, in many instances farmers would make their own trails off of the main roads which were often blocked with solid snow. There were the frozen blueberry sandwiches for our lunches because after we arrived in the mornings our brown bag lunches were put in an unheated cloakroom. At noon we ate the sandwiches with a cup of hot cocoa. The cocoa treat was an organized program to which everyone contributed. Water was heated in two five gallon pots in the one room teacherage, and brought into the classroom at noon. A large jar of good heavy farm cream was supplied by each family on a rotation basis. Occasionally the person whose turn it was to bring the cream would forget, so you can imagine how the cocoa tasted those days. Firing the schools wood heater early each morning paid five cents a day. That meant arriving at school one hour earlier than usual, a task that was normally given to the senior boys in our grade 1 to grade 8 room school. In many cases the grade 1 students could not even speak English. Ukrainian was the common language and outside of the classroom, not a word of English was spoken. I do recall that at least three of the local children who were from English families also became fluent in speaking Ukrainian. Sports took place outside, a gym being an unknown luxury. Inter-school sports, mainly softball, meant a mile walk to another one room school at Side Hill. The walk back from there was four and half miles home after the game which was usually followed by a number of farm chores. If and when there were special events during the year, families were assigned to bring cakes and cookies which was also such a treat. Students having apples for lunch was common during the harvest season, when a box was purchased for the thrashing crews. The rest of the year when someone brought either an apple or an orange for lunch, it was looked upon as having a special gift and each bite was closely monitored by the rest of us. The end of the school year was awaited with great anticipation as it meant a picnic at Sheperts Lake, about one and half miles from the school by and could be reached by talking a short cut through some bush and a farmer's field. This end of year picnic was highlighted by homemade ice cream and hot dogs over an open fire. We would go swimming and play all sorts of games.

Our summer break started July 1 and was the beginning of a hectic schedule around the farm. Mom always planted about an acre of garden. The weeds were now in full growth therefore the weeding process was our first chore. Through these years money was very scarce but everyone worked as a team. We always enjoyed the benefits of having plenty of food, most of it home-grown; beef, pork, chickens and vegetables which were always available. Every week Mom had a five gallon can of cream to sell at the local creamery. It brought in about six dollars. With this money she would buy basics such as sugar, tea, coffee and maybe a treat of some candy and dozen or so apples. The lifestyle of these early years may sound surprising to some of you, but this was common to almost everyone in the whole area at the time. In my family the warmth, the common bond of love was always present and has left the most lasting memory for me.

Great Beaver Hunt

With thoughts of my school days still running through my mind, my mind turned to more immediate things. The last week of April in Embarras Portage was shaping up to be seasonally warm. The heavy snow we received in December and January had begun to melt rapidly. Birds chirped away happily throughout the day heralding the beginning of the new season. Travel along the river by dog team was more present. A lot of the trappers were going to the trading post to barter in their fur catch for food supplies. Creeks in the area had overflows and so they would be running. Pockets of water were common along the river shore line, caused by the runoff. At this time of the year most trappers turned their attention to beavers and muskrats which would bring in the bulk of their annual income. George had talked to me earlier about joining him at a prime beaver location situated on the south-eastern corner of his trap line. He told me about the massive network of beavers that had made their home there. Getting there would not be easy due to the network of creek crossings and difficult terrain that had been covered. George would make the initial trip there by himself, to scout the area and get an idea of what he would need to bring for a successful hunt. Each year George made a number of trips and each session would last four or five days. I made plans to cover off a couple of shifts so I could join him sometime in the middle of May, to take part in this most Canadian of hunts. During the first week of May 1953, very warm weather was part of our forecast. It caused sudden snow melt and local streams to run high. George had made plans to make his first solo trip to the beaver trapping area by dog team. He would take the longer route to bypass a number of creeks in the area. After setting up camp at the spot, he'd leave his dogs chained up and walk back to pick up a "ratting" canoe, which was used to make easy crossings over creeks along the shorter route. This is when I would join him. The ratting canoe is a very useful tool in beaver hunting. I had four days to join George on this unusual trapping experience. He was very excited to have me along help him out. It was difficult to understand the stamina this man had at close to nearly seventy years of age. I was in my early twenties and seemed to have a hard time in keeping up with him. We made preparations for going back to beaver country. It would take us most of the first day to get there. The canoe itself is only some eight feet long and very narrow. The canoeist is required to kneel down when in use which requires very good balance. The ratting canoe is the most useful tool in beaver hunting on open water, for quickly picking up shot animals before they would sink. We would be making a number of creek crossings along this route which meant we would be both in a canoe at the same time. This would require very careful manoeuvring. One bad move would overturn us both into very cold water.

We had to be prepared, heading out into beaver country. I had very warm winter clothing packed in a kit bag, along with some extra wool socks and a sweater. Sleeping outdoors at this time of year was no picnic. I talked to the airport cook about the trip I was going on with George and how I would be gone at least four days. He told me to go into the food store room and pick out what I needed for the trip. I filled a large canvas pack sack with mostly canned goods like pork and beans and stew, also some slabs of bacon, garlic sausage, wieners and other items. I also had a fairly good load of ammunition to pack together with my .22 rifle. When I arrived at Georges home the morning we were leaving, he told me that we had a checklist of items to go over. After a quick look at the grocery items I had brought, mostly canned goods, he told me these items had to be left behind. The only thing he felt was okay was the slabs of bacon, garlic sausage and wieners. George said that on beaver hunts we basically lived off the land. All we needed was tea, sugar, flour, baking powder and lard for bannock and the bacon was for a treat. We would make bannock on a stick and cook over the open fire, and we would trap beaver and roast the meat. Another treat, or so I was told, would be to roast the beaver tail. It was all part of the experience, I thought, and would try it all once with an open mind. Anyways, since we were packing a lot of gear such as traps, ropes, a tarp, guns and food we could not afford the extra weight of the canned goods I packed. I left all that with Granny for the family to enjoy, which was a treat for them.

We said goodbye and left on the journey, heading out into the still early morning light. I'll admit I was unsure about the steady diet of beaver meat, so I sneaked two rolls of garlic sausage, a pack of loose wieners and some hard tack biscuits into my bag. We moved very slowly along a trail, taking turns packing the fairly light canvas ratting canoe on our shoulders. Every so often we would stop for a rest and smoke break. We soon came to a creek crossing about an hour and a half after departure. This creek flowed out of his beaver country area. It was not too wide in places, but fairly swift and high at this time of the year. All our gear was laid down and George said he would transport me first across to the other side. I was leery about getting into this small, very narrow canoe. The only thing I was concerned about was getting dumped in the cold frigid water. George had many years of manoeuvring this canoe so he was confident of getting me across safely as long as I didn't panic, which I assured him I would not do. We were both required to kneel and provide a balance across this fairly swift moving water. It only took a few minutes to conclude my crossing. He then returned a few times more to transport all our gear. Once we finished the trips across we moved on again until we located a suitable place to rest. With all the exercise and fresh air, a break with a couple cups of tea and smoke was an excellent idea. The weather was nice and warm, and the sun beat down. The route along the well used trail seemed alive with many different noises from a variety of birds singing, and the squirrels chattering away in the spruce growths. Three hours into our trip George mentioned we would soon approach another creek crossing. After we get on the other side there was a suitable place for a dinner break marking the halfway point to our destination. The second crossing was accomplished in the same fashion as the first one. I was a lot braver this time. George mentioned that the creek we were crossing was the same one that wound about throughout his line. Once I was on the other side I could see that the campsite he mentioned was frequently used in the past. There was a large log lying down for sitting on. Coals and ashes showed that there were campfires here in the past. I collected some dry branches and a fire was soon blazing away. We packed snow into the tea billy, George threw a couple of handfuls of loose tea into the pot. Granny had packed into George's kit bag some bannock. I pulled out four wieners from my pack to be roasted over the fire. While we rested and smoked after lunch I asked George about hunting and trapping beavers. I was pretty naive about the process between securing the pelts and finally having them sold. George explained that there is a large variation of sizes, quality of furs and methods of preparation for sale. There were five basic sizes on which the furs were graded. Pup, and sizes small, medium, large and blanket. The prices varied substantially between the pups and blanket. He usually tried to avoid trapping the pups, but they are stupid and are trapped more easily. The blanket size ones are very cunning and more difficult to capture. Skinning a beaver also varies in time, this too depending on the size. It could take anywhere between one half to one hour. It however takes a great deal more time in having the animal stretched on a plywood board. The biggest job was to scrape off all the fat once the hide is properly stretched, than nailed down on the outer ends by small nails. Once properly stretched and nailed it takes on a pear shape. In order to calculate the size between the pip and the rest central measurements are taken both in weight and height. Great care must be taken as not to damage the hide during the scraping process. After stretching the beaver pelts are hung up in an area for slow drying. Once drying is completed, they are packed away in a safe place ready for market. The marketplace was always a major issue for trappers to consider. Prices could vary widely from year to year or even month to month. A lot of the local trappers usually depend greatly on the honesty of the local trading post owner. Some ship their furs to the fur auction in Winnipeg, but usually have to wait a long time before receiving payment which wasn't always an option. There are also times when the trappers receive less from the auction than they could have locally. For these and other reason the greatest number of trappers preferred to deal at the local trading post. It gave them the flexibility of maintaining a credit rating for groceries and other essentials. George mentioned that most of the natives prefer to trade locally. Pup belts usually brought in about twenty dollars. He has had blankets coming in as high as one hundred dollars. A good average, however, is in the neighbourhood of fifty dollars.

George's line was home to a fairly large population of beavers, for which he was very grateful. Between the squirrels and beavers this continued to be the Letourneau's main source of income. I listened to what George had to say with interest, leaning back on the fallen log, puffing casually on my home rolled cigarette. The conversation gave me a much better picture of how the fur industry was handled in the north. We repacked our gear and were off down the trail with renewed energy. We would be reaching our destination in the next couple of hours. After being on the difficult trail most of the day, we both started to show signs of getting dragged out. Suddenly I noticed there was a distant noise that sounded like a waterfall. As we stopped to rest George told me that sound meant we were nearing beaver country. The noise came from the beaver dams that criss-crossed the mass of small creeks that entered into a fairly large lake. It wasn't long before we approached this beaver sentry and it was purely a phenomenal site to witness. It was a different world to be in, a world untouched by man that few people had ever seen. Patches of snow still littered the area. As we entered the site we did notice was a great number of wolf tracks in and around George's campsite close to where the dogs were tied down. The five dogs had had been tied up some four days earlier seemed to be in good spirits and were very glad to see us. Two of George's dogs were female and it was the time of the year for their romancing. It happens occasionally that they are bred by the wolf males, but it was unknown if anything happened in this case. The dogs were naturally hungry as the food the George had left at the time of his departure was long gone. George had a gunny sack of dog rations packed and tied down in a tree. They would also be fed beaver meat once he caught some and got them skinned. I surveyed the scene. This was a campsite which he had used from year to year. It was on higher ground near the lake, surrounded by spruce trees. There was a fire pit, and a lean-to construction consisting of poles covered with spruce bows. Near the fire pit and hung on trees was a very used and fire burnt frying pan and a tea billy, which was actually a three pound lard pail. Also in the tree tied down in a bag were a few cooking utensils, tin cups and plates. George classified this as his "hotel". It was getting well into the late afternoon and George's first chore was to check a number of traps he had set earlier in the week. These were set near some beaver runs. While he was doing this I gathered a supply of dry wood and soon had a fire blazing away. George returned some half hour later packing a couple of fair sized beavers. As these animals had been dead in the traps for a number of days they were not good for our meat supply, but would be a real treat for the dogs. In the failing light we sat around the campfire; I made a pot of tea while George was busy skinning these two animals. He would rough skin them and roll the hides up, and then pack those into a safe spot in the toboggan. After the skinning was complete he would as carefully as possible divide their carcasses into five sections, one for each of his dogs who had patiently waited for their turn at the dinner table. As soon the meat was thrown at them they devoured it with big gulps. After their supper meal the dogs laid down to rest. As we sat around the camp fire the roar of the beaver dams continued and we both commented on what a peaceful place this was. There was still some bannock left and roasted garlic sausage over the fire seemed like a good idea. Tomorrow we hoped to have some fresh beaver meat and also mix up a new batch of bannock. We repacked the tea billy with some fresh snow to brew a new batch of tea. There was a great deal of fresh water around but George claimed it was not a good idea to use this water due to number of beavers which would lead to getting ill with "beaver fever". After a hearty meal and a number of cups of strong tea we settled down keeping the camp fire going. Once darkness fell it cooled off considerably. Although I had been prepared for this, the chill literally went to my bones. It looked like it was going to be a long night, with the only shelter being a small tarp hung over the lean-too. Our bed would be a soft pile of spruce bows. We had earlier stacked up a good supply of dry wood to keep the fire going all night long, mainly for warmth but also to ward off any wandering wolves. We would sleep with all the clothing we had on as well as not making changes until we get home. Early in the night, laying on the spruce bows seemed quite comfortable other than the smoke drifting in from the camp fire causing some irritation to my eyes. George and I took turns stoking up the fire, but as the night deepened the air became colder. Both of us slept with our feet nearest the fire so our bodies got cold while our feet were heated up through the rubber gum boots. Soon George was sleeping and snoring away. I, on the other hand, had an extremely hard time trying to get my sleep. In addition to the uneven heat and a sleeping area that was filled with smoke, the wolves were nearby making a lot of howling and strange noises. They were perhaps curious with the fire blazing away, as well as trying to make friends with the female dogs as they too barked occasionally to let them know they were still around. George had his 30-30 loaded and ready for action leaning against a tree near the fire in the event they got too close. It was a long, uncomfortable and above all cold night that would not easily be forgotten.

At daybreak George was up and preparing to go out to the nearby lake as beaver movements were noted to be quite active at this time of the day. George would use his canoe to move about in while I would remain near the shore line. Each one of us had our 22's ready for any action, being careful of the others whereabouts. During this early morning George was able to bag two more beavers, one small and one medium. I was not lucky yet. After this we returned back to camp for breakfast. We had a dry pack of mixed bannock which I mixed up with some boiled water. I put a small portion of the bacon into the well used frying pan. Once the bacon frying was done, I would use the grease for frying bannock. George gave me some instructions on the dinner preparations while he skinned the two beavers. The best cuts of these beavers, he told me, will be used for our dinner and supper meals. I was not used to the type of meals I was consuming. Even though I enjoyed them fairly well, my stomach was telling me a different story. I was lucky to have taken along a small supply of Tums which seemed to stabilize things fairly well. After breakfast we returned to the hunting grounds, hoping to cash in on some additional targets. George was busy setting extra traps so he told me to set myself up near a large beaver house and wait. I found a comfortable location and settled in for some action. There was now a warm sun beating down and due to lack of sleep on the previous night I had a hard time to stay awake. It was at least an hour or longer before I noticed a large beaver surface near the house and started proceeding towards a creek shore line. My 22 was ready for action, and taking careful aim through my scope I fired. I did not kill this animal instantly; however, it was hit in the area around his mouth and teeth. He could not dive without take on water, so he proceeds towards the creek bank. The creek had ice with some difficulty I tracked the animal, he had crawled into this crevice and stayed there dying from its wounds. Once I had the animal dragged out I could hardly believe its size. I had a difficult time lifting and carrying it properly. It was in the 70-80 pound range, a real big one. I still had about a quarter of a mile to carry this huge animal. Shortly after getting back to the campsite George also returned with another beaver. When he saw what I had done he was extremely surprised. It was one of the largest he had caught in a long time, an old male who are known to be very cunning and exceptionally difficult to trap. This pelt would definitely be classified as a blanket. We built up the fire and made another pot of tea, George set down to skinning the animals. During our conversation George talked more about his roughly forty years of trapping in this area. He spoke of his love for the freedom this life offered him. He also spoke of his family; Granny and the girls. He loved them dearly but had always wanted a son and although he had met many young fellows from our base over the years, none of them showed the interest in establishing the down to earth type of relationship with him or his family to the extent that I had. These remarks gave me a warm and comfortable feeling, and I knew he was suggesting he wished he had a son like me. We sat like that for a while, watching the fire and enjoying the great outdoors. It was getting late into the afternoon when we again made tracks to see if we could peg a few more of these prize animals. George got back into his canoe to cruise the lake shore, while I returned to a location near another beaver dam. While sitting quietly and waiting, a small bunch of ptarmigan landed nearby. Knowing the meat is very tasty from these birds I quickly took aim and managed to take down two of them before the rest flew away. I was proud of myself for the fact we would have fresh meat other than beaver for our supper. In addition to ptarmigan I did not return empty handed. I managed to score another average sized beaver as he climbed up the nearby bank. George returned with two animals we well, one he shot and one he got in the trap. Darkness was settling in as we stoked the fire and made plans for our supper. George was pleased with the ptarmigan which we prepared for roasting over the fire. We would have them with tea and hardtack biscuits. Later in the evening we would sample beaver tails which would be roasted over the hot coals of the fire. Personally I could not see what a great feast it would be to eat beaver tail, but was willing to try as George claimed it was a tasty delicacy. The tails remained in the hot coals for a considerable length of time. After cooking the rubber outer skin was removed. It was again placed in the coals for a touch up before salt and pepper was applied. There was very little meat on the bone type structure, which appeared very greasy, and my taste buds did not find beaver tail to be very special, but I told George I enjoyed it anyway. Later that night my stomach did make some summersaults which I endured in silence. After supper in the light of the large blazing fire, George skinned the latest beavers and rolled them up in the toboggan. The dogs enjoyed the continuous supply of beaver meat they were receiving and laid around in total comfort. During the day the heat made things comfortable, but night was again approaching so I expected another cold, sleepless night. I made sure we had an adequate supply of fire wood. As expected another cold one was in store. The smoke filled campsite and lean-to made getting any peaceful sleep somewhat difficult. The various sounds of wild animals roaming around nearby didn't make things any easier. The wolves were very close and making conversation with the dogs who hollowed back. One thing was for sure- the large blazing fire kept them at a small distance. I got up often to stoke the fire have a smoke and even brewed a pot of tea. Throughout this entire ordeal George was out like a light sleeping and snoring away.

Nicholas Verbisky

With the arrival of morning the sun poked out of the horizon and the sign was another clear and comfortable day ahead. The snow was disappearing rapidly, with the water flowing into the lake and various creeks near our campsite. George woke up at daybreak again and after having a cup of strong tea that I brewed earlier was off to check his traps. I was left to prepare breakfast. Our regular food supplies were quickly disappearing but we felt there was enough to last us for the remainder of the trip since we were heading back to home base tomorrow. Our menu was very limited, something I was not used too. We would again have some bannock and I would fry it together with a small supply of bacon we still had on hand. This, along with some fried garlic sausage and strong tea should be sufficient to hold us over. It wasn't that long before George returned to camp packing two beavers and stated he had left one behind to be brought in later. He told me he'd trapped all three of them in one central location near to a large beaver house. Life around this camp had become regular routine with one goal in mind: to have as many beavers as we could get to pack home. George had very carefully studied the habits and family structures of the beaver population on his line. He knew the overall totals and would only harvest a certain number, careful not to cause total deprecation which would affect his future trapping income. After breakfast while George was busy skinning it warmed up sufficiently so I could lay down for a rest which I hadn't enjoyed recently. The spruce bows suddenly appeared very comfortable and it was not long before I fell into a deep sleep. George woke me up a couple of hours later and for the first time since we had set up camp I felt rested. Before noon on our final day I hung around camp cleaning things up and getting a good supply of wood. George returned to the lake to check on more traps, also to determine any new movements from the beaver population. As I sat there on fallen spruce tree which was used as our table I recalled the last three days. It was like a dream living in a different world. Other than the sounds of various creatures, the peace and quiet was overwhelming. I could clearly understand George's joy in the life of a trapper. I could not, however, comprehend the fact of being so far away in the bush by yourself. What would happen if one received an injury, such as a

Embarras Portage: Before, and After

broken leg or fell into the water? There would be no one there to help you in that case, a frightening prospect. With the lunch hour approaching I looked over a couple of remaining chunks of beaver meat. George had set these aside to be roasted. I prepared to have them placed over the slow burning fire. I strung them on some metal wire and hung them from a tri-pod. In the traditional way, first I salted and peppered this meat with a great deal of care and attention. The large pile of hot coals would provide enough heat to cook these chunks of meat. George returned soon after and had another beaver in tow. An hour or so later we enjoyed the roasted beaver and more hard tack biscuits and tea. I could not get over the amount of tea we drank out here. It was loose tea made strong with sugar, and it tasted so good. George had studied the beaver habits very closely stating that during spring or summer beaver see very little sunshine. They come out mostly at sunset unless kicked out by the old girl when she has her young ones in the beginning of May and not long after the sun is up in the morning they return to that lodge to remain for the day. Beavers are nocturnal workers, but more than likely these habits are a form of defense from their enemies which are numerous. By the time they begin to chew through the ice in the spring, beaver are well furred from their daily ice-cold swim to their bed. Their pelts have not yet been damaged by fighting with neighbouring clans that will sometimes gets amorous ideas about the female of the lodge. George went on to say that he has seen pelts with more than twenty five holes an inch long made by teeth strong and sharps enough to cut down huge poplars that grow near the water. This part of the beaver season when hunting is good is very short, as little as a week or up to three at the most. It was a time of year he considered glorious. The birds will soon be coming north again, some of them to stay for the summer, like the ducks, geese and the cranes that will nest by lonely lakes and creeks in this wilderness. As he had for so long lived with the birds and animals as companions, he could not help but to think of them as co-dwellers of the earth. After a hearty supper we made our final trip to the beaver dams knowing tomorrow morning we would return back to home base. The sun was moving down to the horizon so it was feeding time. Earlier I noticed some freshly cut poplars on a nearby bank not too far from one of the mud houses. I positioned myself in a comfortable location and waited. George had got into his ratting canoe to cruise the area. We were not in sight of each other, hoping for some additional catches. It wasn't long before I noticed one of our targets surface and gently swim towards the shore and slowly move up the bank. It was an average size animal and as I was in a good position I took careful aim through my scope. I gently pressed the trigger and dropped this poor creature in his tracks. Retrieving it with care, I turned to my original spot for maybe another hit. While I waited I heard shots ring out assuming that George has also been successful with his hunt as well. As I sat quietly waiting two more surfaced and together were swimming towards the feeding area. I prepared myself carefully and as they left the water and started climbing up the bank I again took aim and as before this animal did not have a chance. The second one made a quick turn around and with a large splash disappeared under the water never to surface again. I was very proud of myself for getting these two pelts for George's collection. Darkness was quickly moving in as we both returned to camp. George also had bagged two of his prize animals. We rekindled the slow burning fire and the fresh blaze gave us some light. While I repacked the tea billy with snow, George sharpened his skinning knife and got down to business rough skinning our most recent kills. The best cuts were set aside and they would be packed home for the family to enjoy. The trip back tomorrow would be a hard pull for the dog team so this extra feed would give them the necessary energy. After the skinning was complete the pelts were turned fur side out and packed into the toboggan. With darkness now again upon us we again prepared to catch some shut eye. There was again the sound of various creatures all around us. As clouds had moved in earlier in the day, a very light snow was beginning to fall. The wolves once again began to howl, their baying returned by the barking of the dogs which went on for quite some time into the night. Like the last two nights, I did not sleep well although it was not as cold as previously experienced. Stoking up the fire frequently I was able to drift off for some rest. George as usual snored away peacefully. Daybreak was early and we would be making preparation for our trip home which I was kind of looking forward to. In the morning we accompanied George back to the beaver dams to check the traps set under the water near the entrance to their lodges. We would also be lifting the traps for the return trip. A total of ten traps were removed and four more animals were caught. We had some heavy packing to contend with back at the camp. I was the camp cook once again while George carried out his skinning duties. We would be required to eat a substantial meal to fuel us for the long day ahead. Other than the beaver meat our supplies had been mainly depleted. For breakfast we would have freshly mixed bannock with the remaining garlic sausage. There were still some wieners we would have for our noon meal. We seemed to have an everlasting supply of tea and sugar to keep our energy up. The dogs were fed early in the morning and appeared quite frisky, like they knew they would be heading home as well. Breaking camp meant making sure the fire was totally out, hanging some of the items on the trees and backing the toboggan carefully. There was over twenty raw beaver pelts and a good supply of choice beaver meat. Add to this our personal supplies and the toboggan was almost fully loaded. A tarp was placed over the whole sleigh and carefully tied down. Our next chore was to get the dogs ready for the trip. The four dogs were organized and arranged in proper positions. The two main dogs on the team were the lead, which was the head of the pack and provided the direction which the teamster wishes to proceed. The wheel dog is the last and nearer to the toboggan. He is the strongest and provided the greatest position for pulling the load. The two dogs in between are followers and basically provide and assist with the pulling power. The lead dog is usually the most energetic and easily learns to take directions. If the teamster wishes to go the right he hollers out "GEE". If he wishes to proceed to the left his cal is "HAA". As usual at the beginning of the trip, the team took off with a great deal of speed. I hung on the top of the toboggan, while George stood behind. We would be proceeding in kind of a roundabout way because a lot of snow had melted and there were many ground spots. The trip would take us away from the creek, and we would follow an old trail which was used mainly during the winter. The packed snow would make the pull of the sleigh easier. As we moved along at a fairly good pace, my mind drifted up to the last almost four days away from home. My eyes had been opened up to the carefree and difficult life of a trapper, rewarding in its freedom but not always in comfort. The general hygiene left something to be desired, as taking a bath was a non- existent luxury. I did take along a small bar of soap and towel. We would frequently wash before meals rubbing some snow in our face and hands, then lather with soap and again rinse with a fresh application of snow. This would naturally wake you up in a hurry. I hadn't showered or brushed my teeth since the beginning of the trip. My clothing was badly soiled and the smoke they were exposed to was something left to be desired in terms of smell. A couple of miles down the road, the dogs slowed down to a walk. I removed myself from the toboggan and started quickly walking behind the team. The trail was exposed to the sun in open areas, and there were a large number of ground spots, which made the sleigh pull difficult. There were frequent stops to give the dogs a rest, while we would enjoy a roll-your-own. As we moved into the late part of the morning the sun was belting down, causing the little remaining snow to melt quickly. There appeared to be a great deal of ground showing as well as puddles of water all around. We trudged along. Around noon we located an area of high ground near a large growth of spruce. We would stop here for lunch, the dogs would be very happy with an extended rest break. While we made a fire, the chatter of the squirrels in the spruce trees nearby indicated they were on the move and sounds suggested there is a considerable number around. For our dinner meal we cleaned up the remaining bannock, three wieners each and strong tea with sugar. After a smoke we decided to grab our 22's and enter the spruce growth with a view of adding to our hunting collection. The dogs lay there peacefully but were secured in the event they decided to move. Within half an hour we managed to pick off at least a dozen of the small critters which we packed into one of our canvas knap socks. George stated we were now over half way to our home so we decided to once again move out. After this extended rest the dogs were again frisky and ready to hit the road. To slow things down for awhile I again climbed on the toboggan and we were off in a flash. A few miles down the

Embarras Portage: Before, and After

path the dogs had their tongues hanging out and once again slowed down to a walk. I jumped off the sleigh to ease on the load. We seemed to be hitting a large number of ground spots. We stopped often to give the dogs a rest and enjoy a smoke. It was still well before the supper hour when we arrived at George's residence. The whole family came out to greet us. Edna and the girls took charge of getting the tired out dogs to their kennels. We slowly unloaded the cargo, some to the small warehouse and the furs to the house. Granny would help George scrape and stretch the beaver pelts. When this was concluded I gathered some of my items and headed home in time for supper. First I needed a good cleaning. As I was working the day shift tomorrow I would be able to hit the sack for a well needed rest.

The Mighty Athabasca

In mid May the Athabasca River was still frozen, however there were signs of a possible break up coming soon. All around were small streams of water runoff from the fast melting snow draining into the river. The snow on the great river itself had already melted down. Overflow completely covered the river bed and "rotten" ice was forming. Spring breakup would soon occur, according to the trappers estimate. I watched the river freeze up in the fall, but the break up again in the spring would be a new experience. This would be a great spectacle and everyone on the base was anxious for the miracle to happen. There was no entrance fee for the greatest show on earth. This beautiful river lay right below us so we would have a front seat. On the morning of a day shift, I was having breakfast when we first heard it. There came the sound like the wind in the trees but not grating. There was a mad rush to the river bank to view this event. It was moving ice, now it really was. This was real ice, big blocks and small ones that ground against the shore and against each other and filling all the eddies and dead water places first. The sound was the ice piling high in ridges and the rise of river water was obvious. This event went on for a number of days, sometime stopping due to ice jams further down the river. Eventually there was a normal water flow with only occasional chunks of ice floating by. Large chunks of ice remained on the river banks still a few weeks later. While standing high on the river bank near our radio shack, one clearly noticed the quickly increasing level of the river water, at times between ten and twenty feet above normal. We knew the reason was the massive pile up of ice near the willows sixty five miles or so downriver leading into Lake Athabasca. The ice on Lake Athabasca was still very much intact, restricting the water from entering the lake and causing the river water to back up and cause flooding conditions. With the flow of very high and fast water comes a variety items washed off the banks. There came floating down the current a variety of empty gas drums, lumber of all sizes and amounts, large trees with stumps ripped off that had fallen into the river, an odd boat broken loose and even small buildings which had fallen prey to a loose river bank. As I stood there viewing God's spectacle, I thought about the geographic nature of the mighty and beautiful Athabasca River.

All this has been going on for hundreds of years. The river rises in the glaciers of Alberta's Jasper National Park, one of the most scenic areas of the Canadian Rocky Mountains. It is the longest in Alberta winding nine hundred miles through mountains, prairies, forest and muskeg to Lake Athabasca in Wood Buffalo national Park. It is one of the major drainage conduits for western Canada. Here I was, standing only about sixty five miles from the final destination of this magnificent river, which had been an asset and service in so many ways to thousands of people throughout the ages. The most significant of the historical features of the Athabasca River is the Grand Rapids. These rapids are located on the river between Athabasca and Fort McMurray. The rapids are found by huge boulders that stretch across the river from bank to bank creating a major transportation hazard and a striking geological feature. During the fur trade, the Athabasca River became the Hudson's Bay Company's main river route to the north- and the notorious Grand Rapids were the greatest obstacle on it. The rapids are still remote today and can be approached only by boat or aircraft. The isolation of this area and the difficulty of the rapids mean a canoe tour should be attempted only by the most experienced canoeists. The Grand Rapids are impassable at any water level and have claimed the lives of many unsuspecting or ill prepared canoeists. Traveling the historic boat route down the Athabasca River from Athabasca (formerly called Athabasca Landing) to the rapids, one eventually reaches the tiny settlement of Pelican Portage about forty five miles above the rapids. It was once a small settlement, now completely deserted other than for a few trappers. The first sign of the Grand Rapids comes about three miles upstream where you can begin to hear the roar of it. One hundred and fifty miles downstream of Athabasca Landing, a large island separates the river into an east channel and more dangerous west channel. The rapids run for about one mile, dropping over eleven meters in the first half mile over a sloping boulder dam. Deadly and beautiful, the Grand Rapids have fascinated countless explorers though history. If only the river could talk, what great stories it could tell. As I look over the beautiful river with small pieces of ice still floating by, one knows very soon the waterways will be busy with floating traffic. The Northern Transportation Company Ltd. will have their tug boats pushing barges from Fort McMurray as far as the willows. These barges will be stored until the ice is cleared enough to cross Lake Athabasca to the point of entrance of the Slave River. Tug boats will move the loaded barges down then through a portage at Fort Smith in the Northwest Territories via the Mackenzie River. In addition to the tug boats local trappers will be heading to the Portage Trading Post to sell their furs and get new supplies. Pondering the geography of this waterway, one could wonder what changes would have occurred if the Grand Rapids were not a part of the water system. Surely the town of Athabasca, earlier called Athabasca Landing would have been a major water transportation headquarters serving all points north of it. No doubt it would also have grown into a major centre and a very important part of the Alberta economy. The alternative was to ship supplies via northern Alberta Railways to Fort McMurray which was much more costly and time consuming.

Work at the office after spring break up was routine, but much more enjoyable with the warm weather which meant much more comfort in our office and our living quarters. There was much more activity on the river to keep us entertained too, watching as tug boats moved about both directions and each passing was announced by blowing their fog horns. Joe Hoover, one of three operators left our base after passing an exam for enrolment into the Air Traffic Control training centre in Ottawa. Joe had been part of our staff for a couple of years. He was always interested in upgrading and spent a lot of his spare time take a variety of courses. His original goal was to go to university, but having an interest in air traffic control he decided to continue to receive training with regular pay. Our office was short staffed for only a couple of weeks when we received word that another operator by the name of Lloyd Clarke was being transferred to our staff on a short time basis. Lloyd was coming out of Coppermine, Northwest Territories after being there for a couple of years. He was due to be assigned a southern station, but lucky for us no openings currently existed. Lloyd enjoyed a brief visit with his family before arriving at Embarras. Being experienced with northern duties Lloyd quickly adjusted to the lifestyle of our base. Lloyd was a white shirt and ties type of a fellow. Other than his work, Lloyd was basically a bookworm and was well satisfied with the slow and quiet routine at our base. Like all the new arriving staff, Lloyd was introduced to the Letourneau family. It wasn't long before Lloyd took a shine to Rose. She also seemed to have a special attraction to Lloyd. We soon learned what was going on but thought it was just a flying fancy. It was not. A close relationship began to take shape. Lloyd would put on a white shirt and tie and visit with Rose as often as he could. Everybody else at the base wore jeans and sweat shirts. Wearing a tie was not heard of, although most of us had at least one in our limited wardrobe. One evening Lloyd went to visit Rose all dressed up in a shirt and tie. Just to bug him a few of us also dressed up in our Sunday best, sporting a shirt and tie which brought a lot of laughs at the Letourneau home. In any event Rose and Lloyd became serious with each other and it was only a month or so later that Lloyd announced they were getting engaged. Just as this was happening a serious turn of events took place. A wireless message was received from headquarters that that Lloyd was being transferred to a southern posting to an airbase in Red Deer. This was very disappointing news to Rose and Lloyd. He claimed this would not deter their plans of engagement and their marriage was still very much going to happen. Lloyd departed Embarras with a lot of sadness. Rose and the Letourneau family would miss him, as well as our staff who had developed a liking to this easy going fellow, but we were sure we would hear from him as soon as he was settled in. Not to anybody's surprise it was not even a month later when I received a registered parcel. It was small with Lloyds address in Red Deer as the postmark. Upon opening the parcel, there before me was a diamond ring. Inside were also instructions which stated this was an engagement ring to be delivery to Rose as soon as possible. A sealed envelope also addressed to Rose was enclosed. To comply with Lloyd's request, I delivered these prizes to Rose at the earliest opportunity. You should have seen the joy in Rose's face when I made the delivery. I am sure that she had feared that once Lloyd left, he would forget to fulfill his promise. Communications between Embarras and Red Deer was very primitive, either by letter or an occasional wireless message. Letters arrived and were sent out on a weekly basis. Lloyd did not lose any time in making plans to be reunited with Rose and her two young children. In less than two months he sent out money and instructions for Rose and the girls to proceed to join him in Red Deer. They would have a small family wedding in his home town. The day arrived for family to proceed to Fort McMurray by boat. From there she would take a C.P.A flight to Edmonton where Lloyd would meet her. It was a sad departure but in many respects a very positive one, particularly for the two girls who were both of school age and would be able to attend a regular classroom. George, Granny and Edna while in McMurray would be able to visit with their relatives and have a brief holiday. The remaining family returned back to their trap line home a couple of weeks later. With Rose and the girls gone, the place did not have the usual activity and you could see in Granny's eyes how greatly she missed them. There was definitely a great void in their lives with Roses departure, but I'm sure they also felt their lifestyle would be much more enriched by the love and care Lloyd would be providing for them, as well as the opportunities afforded to the young family outside of Embarras. Things went on nearly as usual for them. The month of June was almost over, and so was the trapping season. All the furs were dispersed, some to the local trading post and another bunch to the Fur Auction in Winnipeg where George hoped to receive a reasonable payment to carry him over into the next trapping year. There was a short period of rest and relaxation but forest fire seasons was upon us. Already the Forestry Towers had reported an odd blaze in the general area. George and many other local trappers were quickly recruited by the Fort Chipewyan Forestry office to proceed and engage with snuffing out the flames. These men are usually not very well equipped with only piggy back water tanks, axes and shovels. Getting heavy equipment such as cats into the area is very difficult. Often George would be gone for days or even weeks at a time. Wages were only about fifty cents per hour, the days are long and payment is based on sixteen to twenty hour days, seven days a week. But the whole total income helped to provide extra needs for the family. Forest fires destroy trap lines so trappers are motivated to put them out as quickly as possible.

River Fatality

Late one afternoon while I was on duty, one of the NT tug boats going south tied up at our dock. The captain came to our office stating his crew had come across a body on a sand bar twenty miles downstream. They had secured the corpse at the spot and flagged it for easy spotting. We quickly dispatched an urgent wireless message to our office in Fort McMurray to be passed on to the RCMP. In a matter of hours we received a response that Corporal Dogg was leaving McMurray and proceeding to our area to investigate. Darkness had set in by the time Corporal Dogg and another member arrived at our base. Daybreak would come early, so they decided to have a meal and catch a bit of shut eye before starting their investigation. One of the RCMP officers directed a wireless message to G. division in Fort Smith requesting to have the body picked up once it was retrieved, and from there sent to Edmonton for an autopsy. This retrieval would involve sending in the RCMP Norseman to Embarras, returning the body to Fort Smith, and then sending it on the regular RCMP flight by Beachcraft 18 which operated daily between Edmonton and Fort Smith. The two RCMP officers left early the next morning to conclude their gruesome task. They returned a couple of hours later with a metal casket. The Norseman was not too far behind, and shortly after they were off to Fort Smith. Corporal Dogg accompanied the flight to Fort Smith while the other member returned to Fort McMurray in the government boat. According to the RCMP officer there was no identification on the body. It was a man, perhaps in the mid twenties, but was really hard to state anything more than that due to the condition of the body. He only had a pair of pants on and the pockets were empty other than being full of sand. We later learned who he had been. Most river travelers are advised to register in Athabasca before proceeding down the river. Rita Bunkett, owner and operator of a small trading post there had been handling these records. Her trading post was located near the Athabasca River boat launch. The river travelers are all advised about the extreme hazards at the Grand Rapids. The only way to get around the rapids was by portaging them by means of a well outlined established route. Records showed that two weeks earlier two young German canoeists had registered at the post. Their plans were to canoe down the Athabasca River with a goal of heading all the way to the Arctic. They were strongly advised not to attempt going across the rapids, even though they considered themselves excellent canoeists. The body found was one of the German canoeists. The body of the second person was never found, nor any trace of the canoe and other personal items ever located.

It was July 1953. The weather was usually hot, and the mosquito population now in full force. Allan had his speed boat fully operational and had collected a good supply of gas and oil for the next couple of months which of course was prime boating season in our part of the world. I accompanied him on his river excursion whenever we had time off together. At this time one of the civil aviations staff members was being transferred to a more southern posting. He, too, had a boat and motor for sale and approached me about buying this unit from him. The boat was a Petersburg Fisherman of wood construction. It was a very suitable river boat known for its stability and suitable to carry up to five people. The "kicker" was a Johnson ten H.P. which was a good size motor for that area. I took it for a test run up and down the river and was very impressed. With a light load I could easily do about fifteen miles per hour. It did not take me long to decide to buy this unit from Cliff Tolen for a total price of four hundred dollars. I spent a lot time cleaning and servicing my new boat. I named her the "Embarras Queen" and painted this in large letters on its side with red paint. I ordered a forty five drum of fuel and motor oil to be delivered. In the meantime I would borrow a small supply from Allan. Excited about my latest purchase I decided to take it for a run to the Embarras Portage Trading Post. I asked Edna to join me, since she had been really down since her sister and the children had left. Edna was happy to tag along. With just the two of us in the boat I could easily bring the boat to a plane position when I opened up the motor for full speed. The river water was quite high at this time of the year so I didn't have to worry too much about sand bars. There was some drift wood floating about so I had to be somewhat careful not running into that. In less than an hour we reached our destination and it seemed such a joy to be at the controls of a boat I now owned. The people at the Post took time to look over my purchase and approved of the name too. After visiting with some of the locals trappers, Edna picked up her mail and supplies, and we headed out. We knew that going with the current we made excellent time however going back home would be a different story as we would be bucking the current rated at about four miles per hour. It took us about one hour and twenty minutes to the time we tied up at George's pier. In the months that followed all the way to freeze up I truly enjoyed my "Embarras Queen". Any spare time off I had I would be exploring all of the offbeat areas from the River. There were many small creeks and waterways of the river that proved to be excellent fishing spots. Frequently other staff members would come along with me to visit abandoned trappers cabins along the river. If we were not doing this, we would visit trapper's homes and talk about their unique lifestyle and day to day activities. Back at the station, it seemed the river itself had a story to tell every day. The water levels changed all the time depending on the ice melt in the mountains or frequent heavy rain falls, and the flow of the river spoke of this. On the water it was always busy with all types of travelers. There were the trappers, alone in small canoes or large skiffs which carried complete families. The large diesel tugboats pushing barges and fisherman in their large boats heading to Lake Athabasca. I would notice large amounts of lumber floating down the river, in variety of sizes. From experience I also knew a considerable amount of this wood would be trapped on sand bars. George and other local men could use any amount of this free, floating lumber to improve their home, shed or high cash. Sometimes I would take the boat and with a proper roping system help retrieve any of the trapped wood items and have them towed to George's ramp to be constructed into who knew what. Summertime in Embarras was all about the river, and the exciting things and new people or experiences it always seemed to bring.

Aircraft Emergency and Flight Procedures

The month of July was a hot one; our tar paper radio shack was not a very comfortable place to work in. In addition to the heat of the sun our radio equipment gave off its share as well. This made the regular routine of establishing hourly weather reports and monitoring the airways for aircraft movements somewhat tiresome. I was on a series of day shifts and on the last afternoon over the emergency frequency I was alerted by a "May Day" call. *MAY DAY, MAY DAY, MAYDAY, this is CF-MGA calling Embarras Radio.* I promptly replied *CF-MGA this is Embarras Radio I read you loud and clear, go ahead. Embarras Radio this is CF-MGA Roger.*

The pilot responded *I have an emergency – I am approximately twenty miles west of your station at 10,000 feet one turning and one not – losing attitude request emergency landing.*

CF-MGA Roger – have your position – no reported local traffic – wind speed and direction and altimeter setting given. Runway 5000 gross runway in good condition. Emergency fire equipment is being alerted and will be on standby upon your arrival – land at your own discretion. Civil aviation was alerted by the depression of an emergency switch in our office. While I stood by to assist in any way, I heard the Civil Aviation staff and any available staff from our department proceed to the runway with the fire equipment.

Embarras Radio this is CF-MGA – Roger CF-MGA this is Embarras Radio go ahead – Embarras Radio this is CF-MGA I am approximately five minutes w of your airport, I will proceed to make direct descent for approach on Runway 29 –

CF-MGA this is Embarras Radio – Roger – altimeter setting 2995 – wind 320 degrees at 08 – no reported traffic – land at your own discretion – emergency fire equipment on standby near the runway.

Nicholas Verbisky

In a matter of a few minutes there was a great roar over the airport. Being only a short distance from the runway I was able to take in most of what was going on. The twin engine Anson WWII aircraft bounced a number of times during landing which caused it to blow one of its tires on impact before finally coming to rest on the side, about three quarters of the way down the 5000 feet runway. The fire standby crews were relieved they had no fire to contend with. Three crew members exited the plane and under the circumstances, seemed pleased that things hadn't been worse. The only apparent problem was a blown tire and a malfunctioning starboard motor. The pilot approached our office giving me the necessary information to be directed to the Air Traffic Control Center in Edmonton through our wireless system. As this aircraft was based in Fort Smith and issued a flight plan from there we also informed our Fort Smith office as to the current status of their flight. All three crew members were sent to our office for a briefing. I gathered that their airplane was part of a survey company working out of Edmonton. The aircraft was conducting aerial surveys of the area in and around the Wood Buffalo Park, a grid type of air photography when their aircraft lost an engine and nearly met them with disaster. All three members were very seasonal with non younger than forty years of age. The pilot and co-pilot navigator were ex World War II pilots, flying the same type of a bomber aircraft which apparently was purchased from Government surplus after the war. The photo technician similarly had a military background. The pilot was to notify his company about the difficulty they had encountered. A long and detailed message was sent outlining their needs in order to resume their duties. The pilot suggested they should dispatch an aircraft engineer to check out the motor to determine if it could be repaired or need to be replaced. Because of the blown tire the landing gear would also be properly checked out. It looked like it would take some time before they would be airborn again, so they requested a Beaver Aircraft Charted out of Fort Smith to have them picked up and taken back to their home base. Wireless messages were sent out promptly to Edmonton and Fort Smith. Any return information pertaining to this issue was to be sent on to their office in Fort Smith. A Beaver Charter Aircraft operated by Associated Airways arrived two hours later. The three men boarded the bush plane and were soon off on the flight back to Fort Smith. The following day the survey company flew in one of their other planes via Fort Smith where they picked up the pilot. On board were an aircraft engineer and also a civil aviation aircraft crash specialist. It was unusual to have all this activity in our remote and isolated base. The engineer and specialist both took their time in fully going over the aircraft. After careful assessment it was decided that the landing gear was okay, other than the blown tire that needed replacement. The engineer made a thorough investigation of the motor and was able to do the repairs with the parts he had brought with him. His recommendation was to have the aircraft flown to a hanger in Fort Smith where a full C of A was to be made. The pilot was happy with the results and ran through a great number of tests. He taxied up and down the runway a number of times revving the motors. When he was satisfied with the operational aspect, he radioed our office with a flight plan back to Fort Smith. With him on board was the engineer. He taxied to the end of the runway, made a final run up and lifted up with ease. Not far behind him was the other aircraft carrying the crash specialist. I remember these events clearly, as this type of activity did not happen very often in Embarras.

It is compulsory to file flight plans for all aircraft movements according to the Air Transport Regulations. This is a protection to pilots who for one reason or another are unable to reach their destination after departure. Flight plans are either filed in two different categories. VFR (visual flight rule) or IFR (Instrument flight rules.) VFR in most instances applied to smaller type of aircraft who do not have the necessary instruments or capabilities of classifying for IFR conditions. IFR flights require prior clearance by the Air Traffic Control Systems. Weather conditions play a very important part in both instances. Flight plans are usually filed with the aeradio operator, who promptly sends this information to the Air Traffic Control Centre. The flight plan carries very basic information such as identification of aircraft, type of aircraft, air speed, point of departure, flight category, destination time in route, time of departure, estimated time of arrival, and the pilots name. Comments: (e.g.) CF-NAV

Nicholas Verbisky

Cessna 180 130 mph EM (Embarras aeradio) VFR SM (Fort Smith aeradio) 1 + 00, D20302 ETA 2130 Z Smith will phone in arrival time to Fort Smith aeradio station. If a pilot neglects to file a flight plan he could run into a great deal of difficulty if he was unable to reach his destination, got lost, or required an emergency landing. There would be no record or action necessary to proceed with a search. If a plane is overdue on a flight from one point to another, the aeradio station at the arrival point reports this promptly to air traffic control after which a search plan is put into place.

My Pioneer Parents

Whenever I was feeling lonely my mind would drift back to my parents, and early life back on the farm in Stry District near the town of Vilna. My mom had often spoke of the harsh life she experienced after she married Dad on February 12, 1928. After the marriage they moved in with dad's parents, which was the common practice of pioneer families in the area. The shared log house consisted of two rooms; each twelve by twelve in size, one making up the kitchen area, the other the bedroom. It was cramped quarters for four adults to share. There was very little privacy, especially for the newlyweds. In the spring Mom and Dad could set up a sleeping area in a vacant grain storage bin which was something they looked forward to.

Dad realized it would be difficult to maintain a family on a quarter section of land and was able to purchase a half section of virgin soil being sold by the Indian Affairs Department. It was located fourteen miles east of my grandparent's farm, passing through the Saddle Lake Indian reserve in the St. Brides District. The road across was a wagon trail which wound around areas of higher ground. It was sparsely populated with Indian residences along this route, and a very bouncy three hour trip each way by a team of horses and wagon. Mom and Dad would work the home base farm at Stry but during any spare time they would drive to their newly acquired property in St. Brides. Their first task was to build a home. They constructed a ten foot by twelve foot shack of rough lumber and applied tar rolled roofing on the roof. Mom used sewn together flour sacks stuffed with hay for a mattress on a wooden bunk bed. Apple boxes for a cupboard, but they did have a reasonably good small cook stove. Mom told us although everything was very primitive, she loved to have her own space. This land when purchased had no broken land, it was all bush, with the exception of a few small areas of prairie.

During the first year they were able to clear about six acres and have the land broken by a local pioneer. Their shack was located at the extreme west end of their land on the border of the Saddle Lake Indian Reserve. The east end of Saddle Lake was less than half mile away. It was a beautiful lake, with white sand beaches approximately three miles in length and about half a mile wide. The main reason for setting up their residence at the west end of the property was because of the water supply, both for themselves and the horses. Most of the natives also built their log homes near and around the lake for the same reason. Saddle Lake also offered an abundance of whitefish. Mom and Dad soon became friendly with two families on the reserve. Wallis Steinhauer and Daniel Makagos families lived a short distance from each other near the lake. They often dropped by to visit and usually brought some freshly caught whitefish. In return my parents brought back some extra potatoes, eggs and house cured bacon from Stry which they would give to these families. Soon a sort of a barter system was set up which was favourable to everyone. Wallis and Daniel would also help out with the land clearing, rot and rock picking, with payments always in the form of food items with extra money nonexistent. The land was very fertile and Mom soon was able to grow a very productive garden. There was an ample supply for our family and neighbouring native families too. The natives, though food hunters and fisherman had very little concept about gardens. Prior to my dad purchasing the land, the Indian families had a well established wagon trail that ran north and south at the east end of the property. This trail was used by most of the natives at Saddle Lake traveling by horse and wagon to the town of St. Paul about fourteen miles to the south east. Even though there were frequent crossings through the land, my parents did not mind as there were still no fences or any livestock to be concerned with. With no refrigeration facilities, Mom always had to be very mindful when it came to daily food preparations. During the summer when garden produce was readily available it was never a problem. The biggest concern was the meat supply. If a pig was butchered outside of the winter months, meat was fried and preserved with lard sealed in jars. Chickens were also put in jars. Bacon was cured and smoked in a slab smoke house.

It was always a problem to move eggs between Stry and St. Brides on the bumpy wagon trail until someone came up with a very clever idea. Using chop (ground up oats) putting a layer in a large pail, then spacing a row of eggs on this soft cushion, followed by another layer of chop until the pail was full the eggs could travel any distance securely. The young family could also depend on wild game such as chickens if fresh meat became a problem. Mom and Dad worked extremely hard and long hours to maintain home base at Stry and their home in St. Brides. Nearly all farm operations were carried out by the use of horses. Dad had eight, although most operations required four including to operate the gang plow (2 for the furrow), disks, seed drill or harrows. The spare horses were necessary, to trade off because of the hard work they were required to perform. Dad was young and energetic, and it was always his dream to own a tractor which would minimize the very hard work and also allow him to break new land. During the late fall and winter months when most farm operations were at a standstill, Dad would go out and seek employment nearby in construction and bush camps at the saw mills. By doing this, he was able to save some money for a down payment on a 15-30 McCormack Tractor. This purchase was made in the spring of 1930. The tractor was his pride and joy, although a very cumbersome piece of machinery. It had steel wheels with clits which was rough riding especially on non-cultivated land. With the tractor dad also secured a breaking plow, a large single furrow unit of heavy steel construction. Dad was able to open up and break more land. In addition to breaking new land for himself, he also broke land for many of his neighbours and by doing this was able to slowly pay off his debt on the tractor. It was during this period that the Great Depression had arrived creating even more hardships as income from farm produce was at the lowest level ever. Our family had to basically live off the land. In 1935 with the Depression easing Mom and Dad with their now growing family decided to move to their farm in St. Brides full time. A considerable amount of land had been broken and a two room log house was built on the east end of the two quarters of land bordering the north south road allowance. Over the years my parents had established a small herd of milk cows and sold cream.

The cream money was used to provide the families basic groceries and other needs. A well was dug near the house to save hauling water from the lake. The new land was very productive and the farming operation began to prosper. Working from sun up to sun down was a daily routine, but Mom enjoyed it all having a place to call her own. It was with great sadness that a return to Stry from St. Brides was made in 1939. The main reason was the long distance to school for the children, in this case for my older brother Stan and myself. It required a daily walk of three and half miles each way, and we were only in grades 1 and 2. The grandparents arranged for a new house to be built in their yard. With a family of now four children it was no longer possible to share the two bedroom house of my grandparents. There was Stanley (b.1929) myself (Nick b.1931) Olga (b. 1933) and Michael (b.1935). Mom never slowed down with carrying out any of her farming chores even though she had four children to care for. The Stry School was only one and half miles away which was much more manageable. Even though we relocated back to Stry, my parents had no choice but to keep farming the St. Brides land as well. Due to the size of the growing extended family there was not enough income generated to live off the farm at Stry alone. It was always a struggle to move farm machinery back and forth between the two properties, a distance of fourteen miles across an old Indian wagon trail. At first dad would move horse drawn machinery such as a plow, see drill, discs or binder behind a horse drawn wagon which would cause excessive wear and tear on each piece of machinery. As time went on Dad was able to purchase a rubber tire wagon. My father was always very innovative in finding ways of making things easier. On the rubber tire wagon he built a wooden platform that would accommodate any particular piece of machinery on it. It meant a much smoother ride between sites and no longer caused the wear and tear on the expensive equipment. At each location Dad also built a ramp which made loading and unloading a simple chore. Once he purchased the 15-30 McCormick tractor the work load on the horses was less demanding, and a greater amount of land area could be covered in a shorter period of time, but it also meant more expense in fuel costs. Sometimes because of money shortages the horses would be used to carry out a great deal of the workload. Every summer Dad had to go to St. Brides for about two weeks to do the summer follow on the land. With school being out and being about ten I was asked to go as well. My older brother Stanley was not too keen on going as he preferred to stay around home. While Dad worked the fields I would be his cook and helper around the house. Mom also planted a large garden on a plot of very productive soil near the house. My job was to weed and hoe this garden as well. Dad was always happy to have his meals cooked by me. Early in life I learned how to cook a nice potato soup, which remains a good standby recipe. It was made of lots of chopped up potatoes and seasoned with onions. Another staple item was pan fried potatoes and boiled eggs, served with lettuce and green onions from the garden. With no power, keeping some of the perishable items was a problem. We managed to keep some items without spoiling by placing them in a pail and lowering them down the well. Mom would send down some jars of canned chicken or lard jar sealed pork as well. On Saturday afternoon of the first week we would travel back home. Sunday was always a day of rest, a closely followed family tradition particularly with the grandparents. On Monday morning we would again load our supplies for another week. Hopefully by the end of the second week we would be able to return home to stay. With Dad working in the fields I stayed around the house, and things got kind of lonely at the fairly isolated spot. I would go by a bush trail to our Aunt's home about half a mile north to play with some of my cousins who were the same age. There were times I was having so much fun I would forget to get home early enough to get supper ready. Dad did not mind too much, as my Aunt Mrs. Shepert would usually send something home for us to eat. By the time I was about eleven or twelve there was a greater degree of responsibility placed on me. Mom and Dad were good teachers, teaching us how to care for the horses first. They broke us slowly in to working the fields, usually starting off with rock picking or harrowing. Both these jobs were not that appealing. Can you imagine walking behind the harrows being pulled by four horses for six hours or more, with few breaks in between, and the dust generated by harrows causing one to look like you came out of a dust storm? I usually walked bare footed and not always able to avoid kicking a rock or two, so you can imagine how my feet looked after a day in the field. One always felt very grown up for being able to help though. Our parents were grateful for the help they received from me and my older brother Stan in those years. Dad slowly showed me the operational part of our tractor as I got older. He eased me into working first the nice level areas of the fields. Disking or tilling were two areas I was able to handle. Dad always worked with horses nearby so if I ran into any problems he would be close at hand. Being twelve years old able to operate a tractor made me feel very important and grown up. Dad had purchased a threshing machine a few years prior which meant big changes on our farm. The harvesting entailed six teams, with hay racks to properly maintain the threshing machine. Very few farmers had threshing equipment and my Dad would work out plans with these neighbours. Each one would help out by providing a hayrack and team. They would thresh at various locations depending on the condition of the stoked grain moving from neighbouring farm to another. To balance the operation, dad was also required to provide a team. I was thirteen years old and my brother Stan who was fifteen worked together to provide this help. Working from 7 to 7 daily was a hectic pace for a thirteen year old. The first week of pitching bundles was the hardest. Completing the harvesting operations each year lasted about twenty days. The only time there were any breaks was in the event of wet weather or broken down machinery. As threshing operations usually carried over until late September, I was required to miss some school. This was very common for the kids in my age group, the teachers understood this and made provisions for us to catch up once we returned to class.

After reminiscing I snapped back to reality, to my barracks in lonely Embarras Portage. The hectic pace I once endured now seemed like a dream. I always clearly remembered my mothers words- you must stay in school and someday do something different other then suffering the hardship of a farm life.

Late July and early August was busy at the airport. The river always had ongoing traffic, with tug boats pushing barges as far as the

Embarras Portage: Before, and After

willows at the mouth of Lake Athabasca. They would return empty and again be supplied at Fort McMurray. From the willows, other tug boats would take over and push barges across Lake Athabasca to Uranium City or Fort Fritzgerald (Fort Smith). The air still showed signs of lingering forest fires somewhere not so distant. The mosquitoes, horseflies and noseeums were a major irritation. With fall now in sight there seemed to be a decrease in these pests, but it was an ongoing battle. Sprays provided to us were only moderately effective. Work at the station was routine, other than an occasional happening which would cause some excitement and create a break in the boredom. Soon I would have a chance to do something truly exciting that I would have never expected in that remote place.

Staking Uranium Claims

It was late summer, and daily wireless communications were carried out between various stations throughout the north. The messages mostly consisted of official government business. During periods when there was not much going on we would chat on the C.W. circuits. We would catch up on the news, unusual happenings or staff transfers. It was a day somewhere around the middle of September when our circuits were alerted about uranium discovery in the Fort Chipewyan area. This news traveled like wild fire. Apparently prospectors were descending on the Chip area staking out claims. This news put a bug in our radio staff, trying to figure out how it would be possible to get involved. None of us had any experience in staking claims, but with the center of activity only about seventy five river miles away it was an exciting concept with many us having a dream of striking it rich. Uranium was in high demand at the time, and if the find consisted of a suitable grade ore it could mean early retirement. The radio staff held a meeting to come up with a suitable plan. There were four of us radio operators operating on a twenty four hour basis. Our O.I.C. Charlie did not have a regular shift. He would cover off times in case of staff shortages, illness, etc. The following plan was carried out; two operators would cover twelve hour shifts while two operators would be free to travel to stake claims for one week at a time. On return the other two operators would be free to do the same. In the event of an unusual occurrence Charlie would cover off for us. It was agreed that Allan and I would be the first to proceed to the prospecting site. Mitch Stone and Vern Polson would remain behind and they would follow our footsteps upon our return. There was a lot of excitement about this new adventure. Allan would take his speedboat for the journey, and we quickly started getting our gear ready. From the Civil Aviation Branch we borrowed axes, two small sledge hammers, a bow saw and some other tools. We secured a waterproof grub box and the cook agreed to provide us with food rations for the trip. He was very generous in stocking us mostly with canned goods. Coffee, tea, four dozen eggs and three slabs of bacon, four rolls of sausage were added to the list. We packed a duffle bag with clothing and boots and sleeping bags. Additional jerry cans of gas were also loaded but we knew we would have to purchase more en route. The space in Allan's boat was limited, so we loaded carefully. It was our plan to leave early in the morning, only two days after we received word about the uranium finds. We felt we had an edge on some of the prospectors that would be heading to Fort Chipewyan. We woke up at day break and after a hearty breakfast took off down river at the first light. Due to the load we could not get any great speed. Our first stop was at the Embarras Portage Trading Post. We were their first customers of the day. Gunner the store clerk asked us what we were up too so early in the day. He noticed our loaded down boat and was curious what was on our agenda. We told him about our forth coming adventure, that we no doubt would hit upon a rich uranium claim in Fort Chipewyan. He was excited for us and wished us a successful trip. After purchasing some smokes and confectionary items we headed down river again. Proceeding northbound we enjoyed the morning sunrise. The weather was fairly cool weather but with little wind. It blew noiselessly through the trees which lined the river. We passed a number of trapper's cabins, some coming out to wave. As we were passing the trap line home of Allan Albert, we noticed he was outside and trying to wave us down. Since we knew Allan from previous trips we pulled into shore and tied up to his dock. He invited us in for a cup of tea which was a very welcome gesture. Allan's native wife was busy in the kitchen and had tea and bannock for us. They were both happy about the early morning company and we enjoyed our break and seeing them as well. After a half hour stopover we were on our way. Allan and some of the other local trappers had warned us it was kind of tricky of getting into Fort Chipewyan. There were two ways to get there. One could choose to go across the delta which consisted of mud flats with a creek that was navigable but hard to distinguish because the whole area was covered with only about six inches to a foot of water. One could get stuck in the mud flats and have difficulty getting through, especially with a fully loaded boat. Due to the size of our load we could definitely encounter some problems. The second way was to go to the willows which were the mouth of the Athabasca River and enter Lake Athabasca. It was about twelve miles across to Fort Chipewyan. The difficulty there was the waves which were huge if the wind from the east was blowing. The lake was about one hundred miles in length therefore we understood what could be a problem. Our crossing was in extreme west end of the lake. Since it would be well into the afternoon when we would be making the crossing we decided on first trying the "Galute Creek" route across the delta, and cross our fingers that we would not encounter too much difficulty. It was shortly after noon when we approached "Galute Creek", according to our homemade map given by a local trapper who was very familiar with the area. When we left the river and slowly manoeuvred down the creek it was smooth sailing at first while the creek channel was easy to see. We traveled for a couple of miles zig-zagging along the channel before encountering difficulty. We drifted off the channel and got stuck. There were two paddles in the boat so with a lot of pulling and pushing we were able to free ourselves. The whole delta was a mess of small knolls, but basically was totally covered with water. It was obvious that as the creek channel twisted and turned we would encounter further difficulties. Looking everywhere we seemed to be in the middle of a lake and lost in the wilderness. We had no idea how far we were from Chip and since it was already late in the afternoon we were concerned knowing we were now at a point of no return. Little by little we moved, slowly feeling our way along the channel often becoming stuck or hung up. It was exhausting work and we could see ourselves spending the night in the boat. God, however, was watching over us as we heard the noise of a motor approaching in our direction. Sure enough there appeared a canoe driven by a small motor. The native driver greeted us, and could see we were having serious problems. He was heading for Chip and being local knew the creek channel well. He suggested we closely follow him to which we most humbly agreed. We were two or three miles from Chip and following close behind him we didn't encounter any further problems. On arrival we offered to pay this young man for his troubles but he refused to take anything. We were most thankful for his help and as he was a resident in Chip we asked him where to find a good place we could rent a room, and also if he knew of anyone we could hire to have our gear moved. He suggested the only place in town that rented rooms which was Mah's Café, which had some room upstairs. He also knew of a local who provided taxi and delivery services. Since he was going home he would contact this person to move us , pick up and transport to the Café a distance of about a quarter of a mile from the landing dock.

It was late in the afternoon before were able to get to the Café hoping that he did have room for rent. We were in luck, with only one room left and it did have two single beds. The room was far from the royal suite, but it would serve the purpose. We left our wooden grub box locked and secure on the boat. It'd been some time since we had enjoyed a good Chinese meal so it was a treat for which we paid dearly. The food at Mahs Café was very good if somewhat overpriced. Since we were on a prospecting trip we were considered high rollers with plenty of dough, somewhat ironically. As I rested after supper I reflected on our trip through the delta. It was known as a trapper's paradise because of the thousands of muskrats that made this area their home. A marsh land spread over hundreds of acres, it had many creeks flowing out into the rivers surrounding. The Peace River entered from the west, while the Slave River exists flowing to the north. It seemed like a nightmare for one that was not familiar with the area. In the fall the delta was a stopover for millions of wild geese and other birds getting a good feed before heading south for the winter; an event that also brought many hunters looking for fertile grounds. What a sight those thousands of bird must be then, a living sea of white, grey and brown along the flat wet earth of the marshy delta.

Charlie Mah was the owner of Mah's café and was aware of the prospecting activity taking place in the area. There was already a company in town renting space in one of the trailers. Some locals were also geared up and had gone out to stake claims. We told

Nicholas Verbisky

Charlie we were interested in seeking out a local guide and a helper who was familiar with the area surrounding Fort Chip. He suggested we get a hold of the Fleet family, a big family with a few young men who would probably be interested in taking on this type of work. He directed us to the Fleet house which was only a short distance away. At the door we were greeted by an older man. Entering the house it was very obvious that this was a big family. There was the mother and at least four young girls and three young men sitting around the kitchen area. The family knew we were strangers in the small town. We outlined the purpose of our visit while one of the girls boiled water for tea and asked that we join them. The boys talked with their parents and it was decided that Sonny who was sixteen would be able to join us. Like most people living in Chip, they were very familiar with the surrounding area. The prospecting area in question was six to eight miles north of town, just off the Slave River. Sonny knew the area well. He was a hunter and frequented it often. After some discussion we agreed to pay Sonny twenty dollars a day including meals while on the land. The plan was to leave early the following day. We would be able to go down the Slave River by boat for approximately four miles than follow a creek tributary that should lead up close to the proposed staking site. Allan had made some inquiries pertaining to Quartz Claims from the Department of Mines and Minerals, so we were aware of the procedures we had to follow in recording our claims. In the meantime, we decided to stay and visit with our new guide's family. The Fleet's were a very friendly bunch and we enjoyed their company. They were curious about our expedition and wanted to know all about our trip. We told them we were hoping to spend some time exploring the Chip town site but this have to wait till after we staked our claim which we estimated would take around three days.

Sonny greeted us just after day break while we were having coffee at Mah's. We asked him to join us with the understanding we would prepare breakfast once we reached our site. With our gear intact we asked a fellow to drive us to the dock. Sonny gave us direction and we proceeded down the Slave. Within a half an hour we reached the creek that would take us close to our destination. We reviewed the

Embarras Portage: Before, and After

home made map as well as the Department of Mines and Mineral Map. It was determined that our proposed claims were situated approximately in unsurveyed township 113, range 7, west of the 4th meridian. We pulled our boat into an easy exit spot. It was difficult to understand how the terrain had changed just a few miles from Chip. There was mainly solid formation of rock veneering the whole landscape. Growths of small poplars, birch and willows grew in areas throughout between the rocks or where some soil deposits were in place. We could clearly see difficulties that we would encounter with slashing a location line. Before we did anything though, we decided to prepare a breakfast. With little difficulty we had a fire started on a rock bed. From the grub box we pulled out the slab of bacon, eggs and other items. It wasn't long before we had the coffee water boiling and frying pan heated up. This outdoor camping meal was a great treat which we all enjoyed in the clear, early fall weather.

Our first job would be to establish the location line for our stake. Then we would take a compass bearing and proceed with slashing a clearly identifying line to which our post claim numbers were attached. Each claim lays 1500 feet in length and is identified by post number. From the location line each of my claims lay 1500 feet to the left. I was going to stake five claims, each 1500 x 1500 feet in size. Allan would also stake five claims, but located on the right side of the location line. In addition to the post number each quartz claim also had to have a name. I gave the claim numbers a name of "Noncio", for my girlfriend at the time. Allan called his "Lucky One". Once we had all the issues identified the difficult job of slashing a distinguishable line was in progress. The roller coaster of working up and down of the rock formations, rock crevices, rolling hills and knolls was a tiresome process. We were very lucky to have Sonny helping out. He was a true bush man, hacking away on the underbrush along this so called location line. Every so often we would stop to take both visual and compass bearings. I was in much better shape than Allan, who was a bit overweight and I could see how tired he was getting. We would take frequent breaks, stopping to boil some water for tea, followed by a smoke. The peace and quiet we enjoyed was something to be desired. Occasionally we would hear some activity similar to ours in the distance- other would be prospectors staking their claim in the difficult terrain. There was also a helicopter circulating in the vicinity. Our biggest fear was crossing a location line already established by someone who had staked claims before us. This would void our progress and we would have to re-establish a new line and start over again. As far as we could see we were on the right track and being early in line of prospectors we would be safe from crossing any other previously established claims. After a lot of sweat, scrapes and bruises we welcomed a dinner break. From the boat we had packed a kit bag of a variety of good items like canned stew, canned beans and a garlic sausage roll. Together with tea and bread this made for another excellent meal. The open camp fire enclosed in a rock bed gave off a warm orange glow as we ate. The weather was warm and there were few bugs which was a god sent. After a short rest we were back on the line slashing our way across the rugged terrain. By the late afternoon we saw our way by at least a good half a mile of a cut line. Sonny was still going strong but both Allan and I were at the point of exhaustion. We decided to make it a day and head for Chip.

Back at Mah's Café we cleaned up in our room and went down for supper. Mah's had a strange menu. It included meat and potatoes but the canned vegetables had to be ordered from a shelf. If you ordered peas or corn they would warm it up for you and serve it with the rest of the meal. I recall ordering pork sausages and Allan ordering pork chops then I ordered cream corn while Allan ordered peas. Some tea or coffee and bread came with the meal. A few other prospectors also were eating at Mah's that night and we talked about claim staking with them. They were working their claims in the same general area as we were, encountering similar problems with the lay of the land beings as it was. Everybody was hoping for a rich deposit discovery so we would be on easy street which made the labour intensive work more tolerable. Later that evening Sonny came to the Café to invite us to his family's home for a visit and have tea with them again that night. We agreed and met them at the house only a short distance from Mah's. The house was full of people. There was Sonny, his parents, two brothers and four sisters. We took notice of the girls, nice looking and in the eighteen to twenty five year old range. Being posted in an isolated location such as Embarras Portage for a long time it was nice being in the presence of some good looking females. There was a lot of enjoyable conversation that night. We did leave early, since we were dog tired from a difficult day and ready for our beds.

Early the next morning we were back on deck looking forward to another day of our mission. As usual Sonny arrived at Mah's and after a round of coffee we headed for the boat dock before winding our way back to the claim site. It was still very early and before we commenced with our heavy duty task we would have a hearty breakfast. Sonny got a fire going to get the bacon and eggs frying. With that under our belts and after a smoke we were back on the slashing brigade. Up and down the rugged terrain marked of by red ribbons we slowly progressed, always making sure we did not drift off the course too far. Occasionally we would regroup to keep everything in perspective. By the time noon arrived we seemed to have made considerable progress. The total length of our location line was to be about 7500 feet. We were about halfway there, which we measured by a rough stepping out system. During our afternoon rest Sonny talked about his life in Chip and his family. The Fleets were Métis, of Cree Indian Decent. His family had lived in the Fort Chip area for generations. They owned a trap line some distance from town, trapping in the winter and working at one of the numerous fish camps on Lake Athabasca during the summer months. In the few days that we got to know Sonny we found him to be a very interesting and easy going young fellow. He knew very little about life outside of Chip. From what we seen it was obvious that Sonny was capable of providing for himself even at such a young age. He was a good and aggressive worker and very much in tune with life in the wilderness. He was a great help to us, and we knew we were lucky to have found him. The afternoon work progressed fairly well and by late evening we seemed to have completed at least three quarters of our task. Dragging ourselves by that time, we decided to call it a day. Packing our gear we returned back to Chip for a clean up and rest before supper once again at Mah's. The facilities for a shower at

Mah's were very primitive but we managed. Some of the clothing we had used on the worksite began to show signs of excessive wear and tear. We had brought some other clothing with us but decided to hold off on them until we have completed our project. Based on what we had already completed we should finish off with our mission the following day. Sonny invited us back to visit their family again that evening. We did not refuse and ended up playing cards and visiting again- somewhat charmed with the young ladies of the house, who seemed to enjoy our company as well. We realized that we required some extra gear that we had not brought with us. Due to the rock formation there would be difficult in securing the claim posts into the ground. We would need a crow bar and a good sledge hammer. Sonny's parents had these tools so they agreed to lend them to us. We did bring along a number of foot and half ¼ plywood squares which would be used as markers. We also had in our tool kit some black paint and a small brush. Once the plywood squares are attached to the claim posts identification markers would be painted on them.

    Daybreak seemed to arrive early and hopefully we would complete our claim staking by the late afternoon of our third day. We followed almost an identical routine as we had for the two previous days. Today, however we would let Sonny complete the slashing while we would concentrate on organizing and securing the claim posts into their proper places. Allan and I looked around for suitable posts which could be hammered down into place. There were some scrubby looking tamarack trees in the area which we felt would serve the purpose. It took some time to gather and get them ready by trimming and sharpening them. Due to the length of the location line it was a major chore in getting them spaced every 1500 feet. We moved our so-called camp about half way down the location line and as usual had a restful coffee break before resuming with claim's post mounting. Allan and I started at the beginning of our location line with post number one. As we had anticipated, it was a difficult process digging any kind of hole in the solid rock bed. With a lot of hammering and prying with the steel crow bar we were able to have the post positioned and then hammered in with a sledge hammer. With nails we were able to attach the plywood boards

Embarras Portage: Before, and After

to the post. This was followed by identification by painting on it with quick drying black paint. We included the name of the claims "Nancio", and in Allan's case "Lucky one". An arrow on each board also included the directions of each claim with mine on the left and Allan's on the right. We were quite happy with our accomplishment. Walking down the slashed location line I stepped out as near as I could to determine 1500 feet while Allan carried some of the gear. We than identified the spot where post #2 was to be erected. After the second post it was noon and Sonny was called in, claiming he was nearing the end of the flagged out area which marked the end of the line. After a very hearty lunch, Sonny was back to complete his slashing while we returned to work on claim post #3. Another hour later we were done and started to transfer our gear to claim #4, using the very same procedure as in the three previous cases. We had fewer struggles with setting up claim #4 as there was not as much of rock bed to contend with and a larger clay base in that particular location. As Sonny was finishing his slashing we were not too far behind, stepping out the location line and struggling with gear we had to carry. This was not an easy task in the rugged landscape we had encountered. Return trips had to be made as we could not carry everything in one trip. When we finally reached the claim #5 position we had to sit down and rest, thankful this was the end of the line. The last three days had really worn Allan and I down, and we were thankful not to be full time prospectors. Claim #5 was again a difficult spot to secure due to the rock bed but now we had Sonny to help us out. He was young and strong so it wasn't any big deal for him. An hour or so later we had claim #5 in place making this the conclusion of this part of our journey. With Sonny helping us with carrying the gear we slowly worked our way back along the location line to the parked boat two miles away. As we neared the boat we noticed some activity in the area. After dropping our gear at the boat we went to see what was going on. We walked in the direction where this activity seemed to take place and soon came across a claim post identified as "Bighorn No 1". It ran in the opposite direction to ours. You can imagine how thankful we were it did not cause us any interference with our claims. The weather was holding out well, a bit on the cool side, but pleasantly sunny with an occasional fine weather cumulus floating by. It was well in the afternoon when we arrived by boat back in Fort Chip. We were very happy to return to Mah's. We had not cleaned up properly or showered for a couple days so we felt very bushed. Having completed our big job we had earned the right to finally change clothes. The ones we had been wearing were in very bad state of repair, so they were discarded. Our plan was to spend the evening and next day resting and visiting with a few of the people we had met. Secondly we would like to make a tour in and around Fort Chip. There was Corporal Harold Rutledge of the RCMP and Constable Jack Forgey who stayed over with us in Embarras. Father Donto of the RC Mission also visited with us on a few occasions so we hoped to look him up. We asked Sonny to join us for supper at Mah's after which we would pay him for the exceptional help we had received. We ordered the best Chinese dishes Charlie Mah could provide. Sonny was to receive sixty dollars for the three days he worked for us but because had had done such a great job we gave him a twenty dollar tip. Sonny was delighted with the cash he was given. Allan and I were again invited to the Flett residence that evening. We again enjoyed the card playing, laughter and good company that existed in this home.

As usual the morning seemed to arrive early, but we knew this day would be for rest and leisure. The town of Fort Chipewyan had a population of around 1500. We were going to explore the area and make visits to a few of the people we knew from Embarras. After a hearty breakfast we set out. We started walking along the shoreline bay of Lake Athabasca with a first visit to the Hudsons Bay Company Store. After getting a fresh supply of cigarettes and having a chat with the clerk, we again set off. There was the local administration building which housed staff of the Indian Bands. Sonny had already filled us in about the cross-section of various people who resided here. The was a mixture which included Chipewayan's, Cree's, and Metis who seemed to live in harmony. A small number of whites completed the town population. As we wandered down the dusty road we noted a large number of shacks and trailers occupied by the locals. Dozens of dogs howled away, some running loose and others tied or chained down. Many of the locals were trappers with dog teams they kept near their homes. We were also told there was a "Dog Island" just offshore, about a mile away from Chip where dogs were looked after by caretakers and paid a fee by the owners. We found it a bit scary walking near these dogs not knowing what to expect.

We were glad to reach the RCMP detachment which was on the far end of town. Corporal Harold was aware of the rush on uranium claim activity in the Chip area. He could not comprehend how aeradio staff from Embarras Portage would end up as prospectors in the area. Regardless, both Harold and Jack were pleased to see us. After a coffee break Harold offered to drive us around the town to see the places of interest. He said the town itself was very laid back. Peaceful at times, but could become rough if booze was brought in or with locals making some of their own brews. Most of the locals earned a living by trapping in the winter and working on commercial fishing boats or at one of the numerous fish camps along Lake Athabasca. Many of the residents rarely travelled to the outside unless for medical reasons. There was a basic nursing station serviced by two nurses who dropped in to see us. They were a couple of nice young girls who were also very glad to see someone from the outside. After that we went to see Father Donto at the RC Mission. Father had stayed over with us at our base on a number of occasions and was also pleased to see us.

Fort Chipewyan is a very old settlement dating back to the seventeen hundreds, as it was on a major fur trading route. Some of the sites were very historically interesting, including both the Anglican and RC Mission sites which had been there for generations. The well established Hudsons Bay Post had provided services to the locals for generations, mainly buying and trading furs. A two man army signal station was also present, providing communication with the outside world. We enjoyed being driven around by the Mountie as he detailed the various points of interest. He dropped us off at Mah's around noon but before he left invited us to his home for supper which we agreed to. Harold and Jack were both treated royally whenever their patrol trips brought them through Embarras. We were certain this was payback for the hospitality they had enjoyed at our base. As we would be heading back to Embarras early the next morning, we had to pack everything and prepare for the journey before we left for supper. There was a considerable amount of gear to get back in the boat, including a five gallon jerry can of gas to be purchased at the local supplier. There were only a handful of vehicles in and around Chip due to the very limited number of roads, so Jack Fargey the RCMP Constable graciously agreed to provide transportation to the boat that afternoon. After lunch we sorted our gear and would get our gas on the way.

According to earlier plans Jack arrived about 2:30 that afternoon and we quickly loaded our gear and enroute purchased gas. At the boat dock, Alan quickly checked over his motor to see that it was running properly. We neatly packed in the gear and covered it with a tarp. A local trapper who lived nearby came over to see how we were making out. When we first arrived in Chip we did talk to him about looking over our property. Since nothing had been stolen we dug into our grub box, which still had plenty of supplies, and pulled out some bacon and a large number of canned items and offered them to him. He was very surprised to receive all the supplies and claimed he hadn't done enough to receive all these goods. We told him it was the least we could do for making sure our boat and all its contents were secure. We returned to Mah's for a clean-up and after a short, restful coffee break at the cafe it was time again to wander off to the corporals residence for supper. That was about half a mile away. The RCMP residences were located in a large, fairly new government building. The downstairs housed prison cells, and office and a garage full of equipment and supplies. The upstairs was the home of Harold, his wife Greta and their small daughter. There was also a large enclosed outdoor compound which housed the patrol boat, a smaller riverboat, the Bombardier and miscellaneous supplies the might require in the performance of duties.

Harold greeted us at the door and invited us in. He introduced us to Greta, a tall slim and very attractive blond woman, and to his cute little blond girl. Harold looked like a typical cop, carried his well over six feet frame with grace. He had spent quite a few years in the north and was well adapted to the local environment. He said he loved the lifestyle and freedom of these northern postings. Being invited into their living rooms we chit chatted and offered us some wine which we did not refuse. Greta was busy in the kitchen preparing the meal, and it was quite obvious it would be something very good by the aroma we were enjoying. Jack, the junior officer, was on a room and board basis with Harold's family. He entered the room and joined us for a glass of wine. Apparently Jack had been in Chip for a little over a year and this was his first northern posting. He too was a tall, slim young man who stated he too enjoyed life in Chip. It was obvious that these two men got along really well. Soon supper was being served and it was something very special. There was a choice of fried lake trout or moose steak, served with vegetables, fresh baked bread and apple pie for dessert. It had been awhile since we enjoyed a good home cooked meal and this was it. After a restful visit after this exceptional meal, Jack drove us back to Mah's. We hit the bed early knowing the next day would be a long one.

We were up before daylight broke. After having our personal gear packed we moved down the stair to the cafe for breakfast. To our surprise, Sonny dropped by knowing we would be leaving this morning. It was obvious he'd taken a liking to us even only knowing him for such a short time. We asked him the best way to get down Galute Creek back to the Athabasca River. He explained there were some markers on the creek winding its way through the delta, but they were few and far between. Sonny drew us a rough map indicating there was a "y" at some point that fed off into the lake. In any event, we would first try that route and if it did not materialize, we would take the "y" and cross over to the mouth of the Athabasca. That would mean a crossing of in the neighbourhood of 12 miles. Sonny warned us that in our small boat, we would be wise to take the river route. The lake was well over 100 miles in size and could generate large waves. This was not exactly the news we wanted to hear, knowing the difficulty of the river route. After saying goodbye to Sonny, we were homeward bound. He wished us a safe journey.

Alan carefully guided the boat on our way watching for any markers. Progress was good at first, but it wasn't long after that he got off the beaten path and we were hung up in a mud flat. With our paddles we were able to move back into the channel. With great care we moved on, trying to read the water that flowed down the channel. With a case of bad luck we soon found ourselves in the same predicament. This time though we were more stuck than ever and pushing appeared to do little good. These struggles went on for a couple of hours- being stuck, and unstuck and moving along only a little way at a time. When we finally came across a marker that suggested there was a "y" we faced a choice of carrying on trying to feel our way along Galute Creek, or proceed with making the crossing the mouth of the Athabasca. Alan and I discussed the dilemma and agreed our best choice would be to take our chances and make the crossing. We moved through the y; slowly towards the west end of the lake. We still encountered some rough spots along the way but were able to free ourselves. It was now at least two hours since we left Chip and we had not covered nearly as much ground as we would have liked. We noticed the wind picking up suggesting that we were not too far from the lake.

As we eased our way eastward we could hear the splashing of waves against the shoreline. As we approached the lake all we could see were rolling whitecaps coming in our direction. Looking toward the west there seemed to be no end to the lake. We stopped briefly not knowing what to do. Should we make a return trip to Chip, or attempt the roughly twelve mile crossing? We could clearly see that our speedboat was not built for open water travel. Just as we were pondering this major decision, we noticed a tugboat pushing barges from Chip back to the willows the mouth of the Athabasca. It was perhaps five miles away. We decided to make a run for it with a view of tying off with the larger ship. With little notice, Alan revved the motor and we were heading straight into the choppy waves. I was sitting about halfway down the boat while Alan was towards the back steering. We moved along at a pretty good clip, and the waves appeared to become greater in size. Soon the waves were hitting on the windshield and splashing into the boat. Our vessel rocked and rolled like a cork in the water. Even though we had life jackets on over heavy jackets, the cold water was being splashed inside in large amounts. Before long the boat was taking on a considerable amount of water and was soon penetrating our clothes, chilling us to the bone.

We were still a long way from the tugboat and seemed to be encountering a high degree of difficulty. We were past the point of no return. The constant battering of the waves against the boats windshield was filling the vessel with frigid lake water. All I could do was say my prayers, asking the good Lord to get us safely to the tugboat.

By the grace of God, we finally did reach the tugboat. By then I was hip deep in water. A deckhand saw us and helped us tie up. In my young life there were a number of close calls when it came to dangerous occurrences. None came close to this particular journey. It was the ultimate brush with death for me. As we entered the lower cabin of the tug, wet and half frozen, the Captain greeted us. Shaking his head and saying how lucky we were to have safely reached his boat. We stripped of our soaked clothing and a crewmember gave us some mismatched items while ours were hung to dry in the warm engine room. We had some clothing with us in a kit bag in the grub box, which we hoped did not get wet but we would have to wait until reaching the Willows to find out. The Captain and few crewmembers were a very obliging bunch. We were given hot coffee which was most appreciated and chatted about our nearly disastrous experience.

The Captain was a well seasoned sailor who told me he was 60 years old. He'd worked the boats since he was 16. His name was Albert Auger, the name of his vessel was the "Slave River Express". Albert mentioned he'd seen many tragic cases in his time on these waters. He said the lake was unforgiving when one doesn't use common sense. It was quite clear that was our case. Albert's boat was pushing empty barges to the Willows when he picked us up. From there he would pick up loaded barges and return via lake crossing then down Slave River on to Fort Fitzgerald. This was a stopping point due to the rapids in Fort Smith. The empty barges would be exchanged with the full ones out of Fort McMurray and the cycle continued on. As we went, I could feel the large tug showed signs of rolling in the whitecap waves. It wasn't long before we reached the Willows, a well sheltered area just off the lake in the mouth of the Athabasca River. The riverbank areas were cleared up to allow for barge exchanges. The first mate brought the Slave River Express to a halt after manoeuvring it to a suitable position. It was almost noon and there was a very pleasant aroma wafting out of the ships kitchen. We did not refuse when the Captain invited us to join them for dinner. The hot meal helped to warm us up considerably after almost uncontrollable shaking from the cold and wet experience we had just been through.

After dinner the first task was to empty the water from Alan's boat. We were glad to see the grub box was water proof. We got our kit bag out which contained a dry change of clothes. We thanked the Captain and crew for their hospitality and saving us from almost tragic circumstances. Alan and I got busy bailing out the water and within half an hour we were ready to carry on with the trip. We still had a long way to travel and were hopeful we wouldn't encounter any more difficulties. Now bucking the current our speed was slower and was using more gas. After about an hour, Alan noticed that the gas supply was getting quite low. We hadn't realized how much gas we'd consumed trying to get through the Galute Creek route. In hindsight before leaving Chip, all of our difficulty could have easily been avoided by hiring a local to guide us through the creek route all the way to the Athabasca River. It would have cost about $20.

We knew we would soon be approaching the Jackfish Lake Indian reserve, a small settlement of only five or six families. We would stop there and hopefully have the chance to purchase some gas from them, at least enough to carry us over to the Embarras Trading Post. A large creek to the left took us to the first cabin, located on a bank above and serviced by a small floating dock. We came to a stop near the pier. A young native fellow came walking down the back towards our boat. We inquired about purchasing any gas. He said he didn't have any, but told us to go to the next log house a short distance away and we could likely get some there. Alan restarted the motor and moved down the creek. When we parked at the next house, an older gentleman came down the bank with two children. Our request was the same as the first stop. It was then that Alan had a brainwave. Obviously in need of some extra initiative, he opened up the grub

Embarras Portage: Before, and After

box that still had a large amount of food items. We laid out a variety of bacon, hard tack crackers, cheese and canned goods. The value was in the neighbourhood of about $50. We told him we would exchange all these items for 5 gallons of gas. After a quick survey he agreed, but said he would have to go to the neighbours to get the gas. He told us to wait fifteen minutes or so. Once completing the exchange we were off again, feeling much more secure in our journey.

The remainder of the trip to Embarras Trading Post was slower but without any great difficulty. The weather was cooler and a bit on the windy side. This made the trip fairly uncomfortable as we were still recovering from our ordeal. To say the least, we were glad to finally arrive at the Trading Post for a rest and warm up. We had coffee and a snack with Gunner and Mrs. Reid. A few locals were also there. They were all very interested to hear about our "staking of claims" experience. After a restful hour we topped up our gas supply and were off again. We arrived back at the base in time for supper and the home crew was glad to see us after being gone a whole week. First on the agenda was a hot shower, clean up and change of clothes. During the supper hour we outlined our journey to everyone at the table. The type of adventure we'd had did not appeal to everyone. The two other operators that had talked about going out to stake claims after our return were having second thoughts. Mitch Stone and Vern Polson did not feel comfortable with all the necessary steps they would have had to take. After careful consideration they decided they would pass on this uncertain journey. The five days each of work time we owed them, they would take by going to Fort McMurray for a break away from Embarras.

Another Staff Change

A few days after our return Alan and I started the twelve hour cover shifts we owed the other two fellows. It was kind of nice to get back into a normal routine though we found the twelve hour changeovers somewhat exhausting. We had prepared all the documents for our quartz claim registration with the Department of Mines and Minerals in Edmonton. Those were directed to the mining recorders office for further care and attention. Shortly after we submitted the documents we received "certificates of record" for the mineral claims. Now, all we had to do was wait and see what development- if any- would take place in the region. With all that we had gone through we were hopeful something positive would transpire eventually.

It was soon after that Alan received word regarding a transfer request to a less isolated station. His new posting would be to Fort St. John in northern BC. The move would happen as soon as a replacement officer was assigned and transferred to our station. Alan was pleased with his assignment; Fort St. John was easily accessable by road or air and was classified as a non-isolated location. Alan liked the idea of getting his transfer arranged prior to the river freeze-up so he could have his boat and motor shipped out by barge to Fort McMurray. From there it could be transported by rail to his new posting. I was sad to see him go. Alan and I had become good friends and hung out alot together while stationed there in Embarras.

Alan left on a scheduled McMurray Air services flight. A week later a new operator arrived by the same Norseman aircraft. His name was Jeff Hunter. He had just finished going through the in-service training program in Edmonton. Like many of us, he was excited about this isolated northern posting. As Alan had, I too was feeling the effects of being stationed in such a remote location. After two years, I was feeling it was time for a move to a place where

Embarras Portage: Before, and After

I could enjoy more of a social life. I made a request for a transfer, the reply being that consideration would be given to me when a suitable location became available. Once Jeff received the necessary training for our station and was able to handle a shift on his own, I scheduled my annual leave. The station would be short one person but overtime hours would kick in so no one complained too much. The extra money always came in handy. It was the middle of October and most of the harvesting operations would already be over at my families farm. I usually always tried to take my leave so I could help out with threshing, but with Alan leaving it just didn't work out that fall. Instead I planned to spend two weeks with my family helping out with other farming duties. The rest of the time would be leisurely spent visiting friends and family around home, and in Edmonton where I had gotten to know a number of people as well.

Family Visit

For annual leave I hopped the first regularly scheduled air service from Embarras to Fort McMurray. From there I travelled on a Canadian Pacific airline DC3 to Edmonton. After a bus ride I finally arrived home in Vilna. Seeing my family after another year up north was wonderful. There were many get-togethers and so much to talk about, including the adventures over the past year. Mom's home cooking was a real treat after so many campstyle meals. Dad was still very busy working around the farm, tilling the stubble with his new rubber tire John Deere tractor or hauling in green feed bundles. I helped him with this as much as I could. In the evenings I would take the family 1951 Fargo one ton truck into town to see some of my high school friends who were still around. There were still the weekend dances in one of the halls which were always a great place to meet up with old and new friends. My sister Ollie was now working and living in Edmonton. I went to the city for a couple of days to visit and take in some of the sights there. Stan was working with Imperial Oil around Redwater area. He started with the company shortly after oil was discovered in the region, and was busy with his career. I saw him when he came home on his days off, and we both helped out with whatever chores needed to be done on the farm. My family had deep roots in the area which tied us closely to the land. My dad's father, Anton, loved the land of his adopted country and rarely left his farm until he died in 1946. His wife Mary survived her husband by six years. And then there were us children. Even though four of the oldest children in my family had left the area, there were still four at home. Frank (b. 1941), Marjorie (b. 1943) and the twins Lucille and Loretta (b. 1952.) They helped my parents out as much as they could. As always was and will be, farm life was an ongoing struggle but my Dad had no choice but to continue to farm the land in both St.

Brides and Stry. This operation took a heavy toll on both my parents, and all the equipment, both of which had to move between the two places for more than thirty years to make financial ends meet. As it happened, shortly before my arrival on leave a neighbour, Alex Chimko, had approached my parents indicating that he was trying to sell two quarters of land. Mr. Chimko and his wife had separated. She had moved to Vancouver B.C. and together they owned three quarters and a home place next door, and two more just north of my parents land. The former Mrs. Chimko had applied some pressure on him to sell so she could get her share of the value of the property. Since it was a court order, Mr. Chimko had to comply. His asking price for the two quarters was $10,000. It was well known that no farmers in the area had this kind of money, or was it possible to get a bank loan. However, he still brought the sale up with my parents. My brother Stan and I discussed this issue while I was at home. Both of us being from the old school way of doing things, we sure knew how to save money. We compared notes, and decided that between the two of us we had enough money to purchase this land for our parents who had always been so kind and generous to us. Since Stan was older he'd started working a couple of years before me and was more flush. I knew this meant spending my entire savings, putting my dreams of a fancy new car on hold for a little while. I was happy to do it though, knowing what it meant to Mom and Dad. When Stan and I brought up the plan to them, they were overwhelmed. They realized that if and when the new land was bought it meant no longer having to shuttle back and forth between Stry and St. Brides. Arrangements were made with Mr. Chimko to purchase the two quarters near my parents. My bank account of $4700 was drained, with Stan making up the rest. Everyone was pleased: Mr. Chimko, my parents, and Stan and I because we knew how much easier things would be now for Mom and Dad and the rest of the family. My plans for a new car if and when I was transferred back to civilization would have to wait until I built up additional savings. We told Mom and Dad the land was now for their use. All they had to do was pay the annual taxes as long as they remained on the farm. The St. Brides land was rented out without any difficulty.

Too soon, my annual leave was exhausted. I returned to Embarras following the same route I'd left on. Work was pretty well routine. I missed my friend Allan, but still spent a lot of time at the Letourneau home. George was again busy for another season of trapping. I would head out with him on days off clearing windfalls along the dog route, preparing for winter. The weather cooled off quickly and a good snowfall came soon after. This was a sign that trapping season was coming up full speed ahead. As I had for the past two winters, I planned on helping George any way I could. The outdoor exercise was a great way to stay in shape in the slow winter months. And, it was part of the colour and excitement to break up the everyday routine in lonely, wild and isolated Embarass Portage.

Transfer in Effect

One day a wireless message arrived from regional headquarters. Much to my surprise, my transfer request was to be effective at the earliest possible date. It was not, however, the posting I had been hoping for. The transfer was to Smith River, B.C. due to a critical shortage of aeradio staff there. Smith River was another isolated northern posting. One of the operators I worked with had been previously posted there, so I basically knew what the new place would be like. It was on the Amber Two Airways between Fort Nelson, B.C. and Watson Lake, Yukon. It was an emergency airbase operated by the RCAF and our department. There were barracks assigned to us, and I would still receive the free room and board as well as northern allowance. The general area where the Smith River base is located is known as the South Nahanni region, also known as Headless Valley. It's an extremely rugged area about 7 miles from the Yukon provincial boundary. Heading along the Alaska Hwy., the turn off to the airport was at mile 513 along a winding, mountainous route. It was a 25 mile drive to reach the airport base. This was the only way in. You flew via CPA airlines from Edmonton to Watson Lake, Yukon. From there senior officers from Watson Lake and Smith River would make travel plans. The usual way when moving staff to or from the base was to meet approximately halfway along the Alaska Hwy, where the transfer would occur. The Embarras Airport was home to me for more than 2 years. I was leaving it with a great deal of sadness, and of course I would particularly miss the Letourneau family with whom I had spent so much of my time. I had developed a very close relationship with all of them, especially George. On the day I left, the regular McMurray air services flight arrived to pick me up. I cannot recall the exact date but it was some time in October and with my few possessions packed in bags I departed. As I would not be home with my family for Christmas, I was given a few extra days to spend with them. I journeyed to my new posting hopeful that my tenure there would be a short one. Headquarters had already indicated to me that it should be not longer than one year.

Smith River Bound

I wasn't thrilled with my new posting, but I felt a degree of excitement about the new challenges that lay before me. From what I'd heard from the fellow who was stationed there and transferred, it was a much different life and environment than in Embarras Portage. As I boarded the CPA DC3 aircraft in Edmonton on an early morning departure, I started to feel good about what lay ahead. Our first stop over was an hour and a half later at Grande Prairie. I made a quick stop at the old office to see if any of the same staff was there from when I was posted at GP. Much to my surprise, one of the senior operators Bill Sawnson who had provided me with my initial training was on shift. We had a quick chit chat, and Bill introduced me to a couple of the new guys before I had to rush back to board the aircraft again. The next stopover was in Fort St John in northern B.C. and finally, my destination at Watson Lake in the Yukon. It was well into the afternoon when the aircraft pulled into the parking area. Though it was well into the month of October, 1953, the day was pleasantly cool with clear skies and very little wind to disturb the few trees. I noticed right away the Department pickup truck parked on the tarmac. As I stepped off the plane, a middle aged fellow approached me and introduced himself as the officer in charge at Watson Lake. He said he would be transporting me down the Alaska highway to meet with one of the staffers from Smith River at the halfway point. The road leading to Smith River could be treacherous. Whenever staff came or left that base, the dept. in Watson Lake would meet them halfway. That way, the two departments shared the difficult journey. I told him we'd had dinner on the flight but I could stand for a cup of coffee. We did stop at a roadside cafe where we chatted a bit. I told him about my tenure at Embarras and he was surprised why I was not offered a better point than Smith River. Ed stated he'd been posted at Watson Lake for over three years. He was married, with a family, and they lived in government housing near the base. They were very comfortable by now with the unique aspects of the northern lifestyle. Back on the road, we proceeded eastward down the scenic but primitive Alaska Highway. Many mountains surrounded the road, which curved gently while continuously cutting through the landscape. Up and down we drove through the many hills and valleys. The road surface was very rough, dusty and narrow. Extreme care had to be taken when passing other vehicles, particularly heavy transport trucks which seemed to be fairly common. We drove ahead knowing that around halfway to our destination, I would be met by someone from my new posting for the remainder of the way. We rambled along in a rock and roll fashion, and I enjoyed the wild scenery. The mountain creeks made their way down crevasses, and moose grazed peacefully. After we passed another roadside cafe, this man told me we'd past the point of no return and it wouldn't be long until we met up with our contact. About fifteen minutes later, a red pick-up approached us kicking up dust as it went. I was informed this was my ride. Both vehicles came to a halt on the side of the rough highway. We all exited our vehicles, and I was approached by Jim Bresden, the OC at Smith River. To my surprise, Jim was a young man, only a few years older than me. He was very pleased with my arrival since the station had been short an operator for some time, and I would be alleviating a heavy workload for the remaining staff. I took out my metal suitcase and duffle bag and loaded it into the back of Jim's truck. After a brief friendly conversation we were off again heading in opposite directions.

My first impression of Jim was that he would be a pretty good guy to work for. On the other hand, I'd heard he was a no nonsense type of boss who expected good and effective work habits from everyone he was in charge of. This didn't worry me at all. I knew I wouldn't have any problem in that area, and thought we'd get along just fine. The rest of the drive went by quickly with conversation and shortly we rolled to a stop in Coal River. This was on mile 533 of the Alaska Highway and consisted of a roadside cafe, service station and couple of shack looking motel cabins. It was also the mail center for Smith

River airbase. It was now late in the afternoon so we stopped again for coffee and a light lunch. Jim knew the operators of the lodge well and introduced me as the newest member of his team. They were a middle aged couple named Mary and Tom. A young native girl served us our meal. After picking up a small bag of mail we left. The turn off to the air base was at mile 513. Jim warned me as we made the turn that once we left the highway, the twenty five mile trip to the airbase would be very interesting.

Smith River Airbase

As soon as we left the highway, the road turned into a narrow train. Jim expertly manoeuvred the hair pin turns along this treacherous road, if you could call it that, which seemed to scale the sides of rocky cliffs. In places our speed was reduced to ten miles per hour. There were big ravines followed by a steep climb. I had to agree with Jim that this *was* a very interesting road. In places streams of water poured out of mountain crevices and every so often were supported by make shift culverts across the road. There were areas we could speed up a bit, but it was well over an hour before we approached the airport. In 25 miles we had climbed about a mile or so and much the terrain seemed to have come to a plateau. As we entered the airport property I observed a very neat layout of buildings. Jim briefly outlined the various structures. There were two long log buildings- one housed the kitchen and sleeping quarters for six staff, five operators and a weatherman. It also included a lounge area. The second building ran parallel to the first one and housed the RCAF staff and a canteen. Jim pointed out the few other buildings that graced the airport, including a large power house which provided continuous power to the airport. There was a two story airport tower located a short distance away in good view of the airstrip. There were three married quarters also known as PMQ's for the RCAF staff, a fire hall. Storage facilities were also close. Jim said we would have plenty of time to get a better picture of the whole camp, but first should get acquainted with some of the residents. From what I gathered, there was a great variety of personalities that lived and worked here and was looking forward to seeing them for myself. Supper was waiting for us when we entered the kitchen area. I was greeted by a large number of faces. I knew it was always a curiosity when a new staff member arrives. A flurry of introductions and it was a little overwhelming at first, trying to register so many new faces. I'd heard that the meals prepared by this cook were outstanding. Roger the cook surely made an impression on the first meal I was served and I could see that if I continued to eat that well, I could easily get myself into trouble. After the delicious meal, I thanked the cook and proceeded to the barracks area, where I was being assigned a room. The rooms were not large, about 10' x 10' which included a single bed and small dresser. There was a common washroom, shower and laundry area. After I unpacked my few items of personal gear, I met with the staff for a briefing. We proceeded to the tower where Gil Bouchard was on duty. I was joined by Jim, Doug McKay, and Joe Boudreau. Missing was Bill Andriff who was away on annual leave. After introductions we got down to lay out plans of phasing me into the regular shift. It was decided I would need a few days of in-service training as some of their systems were somewhat different from Embarras. One of the things was to get introduced to the teletype land line procedures, which disseminates and receives weather reports and other information relating to air traffic movements. OIC Jim only worked odd shifts to fill in on days off or as required when there was a shortage. Jim happened to be on days the next day during which time I would get started on my training. I could foresee very little difficulty in getting a handle on the work procedures in Smith River. Leaving Gil to finish off the evening shift we all proceeded to the mess hall for an evening snack, which apparently was a nightly ritual. This included most of the airport people. The cook would lay out sandwiches, cake, pies and other delicacies. This normally should have only been a coffee break. Being tired after a very busy and hectic day I was ready to retire and be well rested for the new experiences that morning would bring.

In the privacy of the room my mind explored this new phase of my journey. From what I could see Smith River differed greatly from Embarras. This base was operated by the RCAF, who showed greater care in the construction of buildings, and in the general lay out of the whole base. The tower which was our place of work was well built and possessed excellent outside vision. It was a far cry from the radio shack we occupied in Embarras. This station also differed being it was operated by two distinct government departments. The RCAF was the main operator and in charge of overall operations. Our department provided all telecommunication services and weather reports for the safe movement of all aircraft flying the airways or in the vicinity and there was a joint agreement between the RCAF and department of Transport in place to that effect. The RCAF was not, however, in any way responsible for our staff as long as we maintained adequate behaviour. As I drifted off to sleep, I wondered about my new posting, and what the days ahead might bring.

Morning appeared to come early with a knock on the door. The main joint washroom was busy with showers, but I managed to find a spot. After cleaning up I found my way to the mess hall which already had a number of fellows at the kitchen table. The cook was dishing out whatever one could wish for breakfast. Bacon, eggs, pancakes, sausages, you name it, it was there. There seemed to be friendly exchanges at the table. Jim and a weatherman by the name of William were there having breakfast. William did weather reports on the day shift only. The radio operators did all the evening and overnight reporting. It was common knowledge the weather man had kind of a plum position, being he worked days with weekends off. It was 7:45 am when we made our way to the aeradio tower where Doug McKay was still on shift. After exchanging greetings Doug signed off the daily log and Jim and I signed on for duty. Our first chore of the days was to compile the eight o'clock weather report, which meant to go out check the Stevenson screen for temperatures and calculate as well as observe sky conditions for cloud amounts and types. As I was used to these procedures, I had little difficulty in concluding this hourly report. The next process I was not acquainted with was to prepare and forward it on the teletype circuit. This meant punching a tape of the report and affixing it on a central box which would automatically transmit the report to all stations on the Amber two airways as well as our Edmonton airport central centre. This procedure would kick in approximately ten minutes past the hour. In return we would receive weather reports of all stations from Snag, Yukon to Edmonton with each station contributing theses reports in the same fashion. At fifteen past each hour and a quarter to the hour we would broadcast the local weather report as well an adjustment one these being Watson Lake Yukon and Fort Nelson B.C. The radio communication equipment in general was very similar to that of Embarras. By the end of the dayshift I was pretty well familiar with all procedures. I agreed to take on a day shift the next day and if I needed assistance Jim would be close at hand. From the tower, Jim, William and I went to the RCAF barracks lounge and wet canteen. The lounge appeared a neat place to relax. The canteen was usually opened after supper and carried bottled beer, some types of liquor and the usual supply of cigarettes and confectionary items. Supper hour was at six so off we went to our barracks to relax and prepare for the feast. The next day my shift went favourably. Occasionally there were things I had to check with Jim. The day shift weather man compiled the weather reports, so all I had to do was get them sent out on the circuit. In the days that followed I adjusted easily to the lifestyle of the base. There were many new faces and personalities to get to know still. I estimated about forty people lived at the airport. It was taking me some time to sort out the various personnel and their functions. There were five of us in the aeradio division plus a weather man doing only day shifts. The RCAF had six regular service men with a few being civilians working for the air force. There was a sergeant in charge of the whole base. As well there were two corporals. The two airman had regular duties one in the power house and the other one in maintenance. The sergeant and two corporals lived in married quarters. All had their wives with them and all had small children. The airman's wives were the only females on the base. The Mess hall consisted of a cook and helper. There was also a janitor that cared for both barracks and general living areas. We, however, were responsible to look after our own bedrooms which included laundry. No doubt about it, this was a male dominated camp and being in the prime of my life, there was a degree of loneliness I felt for a social life. Other than our departments staff it seemed that most the employees with the RCAF were somewhat older. A considerable number of them were former military and ending their careers. It was suggested these men were given preference when seeking employment. I slowly started to sort these fellows out and learned that a number of them were married and separated. With a wet canteen being available, many appeared to have drinking problems as well. The weekend poker game was a ritual which would last to all hours of the night. I loved to get into a game occasionally, but I was used in playing with small change. These fellows preferred higher stakes. A number of them frequently lost their whole month end cheque. I could not see the sense of having to live in isolation and then spend your money in such a foolish and irresponsible manner. Most everybody in our department felt the same way, and if we were off shift, often we would gather around the table and watch these fellows win or lose a large amount of cash with sometimes foolish betting usually fueled by their excessive drinking. The only exception to our young staff was Joe Boudreau. He was a Frenchman who came to the area during the late 40's. Joe was in his early fifties, a short and stubby fellow who spoke with a very distinct French accent. Sometimes pilots would have difficulty in following his instructions or deciphering the information he was presenting to them. Joe was separated from his spouse for many years and constantly spoke about his failed marriage. He had a couple of older children back east, but it would have seemed that they had a very poor relationship with them. He didn't play poker, but loved his beer and would usually tie on a good one on his time off from work. Joe purchased himself a scooter on which he traveled around the base. When drinking Joe could be argumentative and be unreasonable however, Jim made sure that once he was on shift he had to maintain good working habits. A number of the fellows working here were of French descent. Joe had the bad habit of conversing with them in French across the table at meal time. This was considered in bad taste seeing as we could not understand them. To cramp his style, I would start talking in Ukrainian to the only other UK at camp, Bill Markowich. He would soon revert back to speaking in English and so did we.

My Friend Albert

In late November the unmistakable signs of winter made their presence known. Freezing rain, intermittent snow falls and the cold north winds made being outdoors unpleasant. Prior to this I had made a point of exploring the whole area in and around the base. Just a couple hundred yards east of our tower there was a gravel based runway that ran in a north-south direction. It was around five thousand feet in length. I was told it was capable of handling all sorts of aircraft safely. I had different ideas for it. I started both running and walking up and down the runway to keep up my exercise program. I was soon befriended by one of the fellows that worked for the air force. He too had tried to maintain some activity, mainly because of a diabetic condition. His name was Albert Vitche. From him I heard the tale of his unusual background and how he ended up coming to Smith River. Albert was a big man, at least six two and weighing over 200 pounds. He was well built and muscular. He had lived in Vancouver with his parents where his dad was a detective in Vancouver police force. In 1940 just after the beginning of World War II Albert was thirteen years old. Not being able to get along with his parents, he left home and attended an Army recruiting centre. Because of his size he was able to convince the Army Office that he was eighteen years of age. He was accepted into basic training which was followed by commando and sniper training. Alberta was shipped overseas after boot camp. Over the next four years he went through a number of European campaigns. Being trained as a sniper and commando he got himself into some very dangerous situations. Upon his return to Canada at the end of the war in 1945 he was only eighteen years of wage. He claimed he grew up a lot because of his war experiences. Shortly after returning to Vancouver he secured a job as a bouncer at one of the fancy hotels. He did that job for a number of years and had established a secure lifestyle. This all changed when one day a drunken patron came at him with a knife. His commando training kicked in and he gave the man a couple of karate chops, killing him instantly. He was charged, but was acquitted being it was considered self-defense. One of the conditions of the charge was that he must leave Vancouver for six years. This is how he ended up working here in the capacity of a janitor or general handyman. Because of what had happened he had no interest of returning to that type of work anymore. In the last couple of years since his arrival diabetes had taken a tool on his body. This was largely his own fault since he had failed to care and monitor his condition properly. In any event Albert and I became good friends.

Coal River Trip

Winter arrived with a vengeance in the first week of December. Heavy snow had fallen and the temperatures plummeted at times to below 40 F. Strong winds caused a great deal of drifting in the snow banks. The RCAF maintenance crew was kept busy clearing the runway and roads. The deteriorating weather brought few changes to our day to day lives, other than that we spent even more time indoors. There was only a couple of hundred yards between the barracks and the tower, with both places being warm and comfortable. Our department also provided us with plenty of warm winter clothing. After a few days of the onslaught from Mother Nature things seemed to improve although it remained very cold. I had managed to earn some days off by working extra shifts. There was a regular mail run to Coal River which happened twice a week. The civilian driver, James Hopper, asked a couple of us guys if we would be interested in joining him on the run. I agreed to go along. The vehicle he drove was a four wheel drive one ton truck with a covered canopy. The road was covered with snow, but had been cleared by a grader by one of the maintenance staff. The trip was hair-raising through this rugged mountainous terrain. James was an excellent driver and very familiar with the route. He mentioned that he was in the airforce and had served overseas. After the war, he got married and settled down working as a truck driver with a long distance transport company out of Winnipeg. After 5 years the relationship with his wife broke down, and they separated. James moved on, getting this job at Smith River where he had been for the past 4 years. From what I had observed, James loved to drink and was one of the high stakes poker players. Soon we arrived at Coal River and after dropping off the outgoing mail bag and parcels, we picked up the one assigned to the air base. We didn't rush back however, instead sitting around the cafe section of the service station and chatting with Mary and Tom who owned the place and also Josephine, the young native waitress. After a couple of cups of coffee each and some delicious apple pie we were ready to hit the road again. Even though we were only 50 miles from base, the road was a challenge and the last 25 miles especially wretched. It was a couple of hours later we arrived at our destination where some were already waiting to see what mail they might have received.

During our conversation on the long and dangerous ride, James referred to the area as "unforgiving". He had lived and worked in the South Nahanni region for years. The Smith River itself circled around the base in a horseshoe fashion. It was not very wide, but very swift flowing in a southerly direction towards the Liard river. He also mentioned there were a couple of fellows who had lived together in this area for a great number of years. They trapped in the winter and prospected in the summer. Their main home cabin was located about 30 miles as the crow flies near a place called Tobally Mountain. Strangely as it may seem, one of the fellows was named Oscar Mills, an Englishman while the other was Helmut Stringer from Germany. These men showed up at the base probably two or three times a year to resupply with groceries and pick up their mail. One of the times they always came out was at Christmas, a time of year the weather was extremely cold; then, they would hang out at the base for a couple of weeks. The airforce officer in charge was extremely accommodating to these two fellows and appreciated them being present for the festive occasion. When they arrived with their dog teams, they would also bring with them a large supply of lake trout for the mess hall as a token of their appreciation.

Airbase Inspection

With Christmas just around the corner, I was told we could be expecting a crew of airforce officials to be making an inspection trip to the base. Sure enough, this occurred a few days later when a flight plan was received from Whitehorse via our teletype circuit. We immediately dispatched it to the RCAF Officer in charge. I wasn't working but was visiting with Gil who was on duty. The airport manager expecting this visit from his supervisors, and he alerted various sections to spit and polish for the event. The aeradio tower was not part of the inspection process; however we kept our facilities in top of the line condition which was the way our officer in charge, Jim, always expected. Roger the cook would prepare food in that any high class eatery would be proud of, and with the inspection we were sure Roger would put a little extra effort in. The flight plan stated it was a DC3 aircraft with flight time being two hours and twenty minutes and departure time being 9:15 a.m. The airport seemed to be a beehive of activity making sure that everything was in place to satisfy the commanding headquarters staff. Most of our staff was in the tower waiting for this special event. Even though we constantly worked aircraft flying over our airways, we did not have too many landings and take offs to contend with, because Smith River was classified as an emergency airport only, and out of the way from any major centre. The pilot made radio contact advising Gil he was twenty minutes out and started descent from ten thousand feet. He was on a VFR (visual) flight plan and requested landing instructions. Due to the mountainous terrain, the pilot had to be extremely careful in his approach. The positive factors for the landing were the airstrip was on fairly high ground and there were no restrictions from either end of the runway. Gil outlined the weather conditions; *Broken Altocumulus clouds, at about twelve thousand feet,*

Nicholas Verbisky

wind N.W. at 8 mph altimeter setting at 2990, no reported local traffic, runway condition cleared of snow. It was not that much longer after we were heard the roar overhead of an aircraft in the vicinity. Then another call from the pilot- *inbound for runway 29 and gear down and locked.* Gil replied *Roger and to land at your own discretion.* The pilot guided the DC3 to a nice smooth landing with plenty of the room still left on the runway. *Time of arrival at 48.* This time was immediately dispatched to both air traffic control in Edmonton and Whitehorse. As the aircraft came to a halt the local air force staff was there to meet them. Greetings were exchanged which was followed by everyone proceeding to the mess hall. As suspected the dining area was set up especially for the visit in a very decorative military fashion. The six inspections personal which included the pilot and co-pilot were seated along with the local air force staff. We took our places at the table as well and after a brief exchange of greetings the meal was served. There was no question in anyone's mind that the cook and his flunkey had done an excellent job. The inspection team departed to various sites on the base which lasted a couple of hours. In the meantime the pilot and co-pilot attended our tower for a weather briefing. These crew members were a couple of fairly young fellows who sat around with us and chatted. They checked on the general forecast and weather back to Whitehorse. This was followed by filing a flight plan back to base. It was soon after that the inspecting officers had completed their yearly task and were driven to the aircraft for departure. Each of the two motors started up, and separately roared into life. The pilot requested take off information. Gil radioed back providing him with an altimeter setting and wind direction plus speed, also stating no reported local aircraft traffic. The pilot taxied down the runway. At the end of the airstrip he concluded a checklist run-up and with a great roar of the engines he was off. Radio call again broke in stating *air force flight 930 off at 17 estimating Whitehorse 2 plus 20 climbing to 9 thousand VFR.* This information was promptly dispatched to Edmonton and Whitehorse air traffic control.

Christmas At Home and Smith River

Being away from my family for the last three years at Christmas definitely made me feel lonely. At this time of year, in the privacy of my room, my mind would always drift back to the joy of being around home during the festive season. Christmas Eve was always a busy time, and included the search for a suitable Christmas tree by my siblings and I at one of the few spruce tree sites. On this day we would also fast according to our religious beliefs. Supper would consist of meatless and fat free foods. Mom, however, knew how to prepare a very delicious meal mainly according to Ukrainian tradition. Then at about eleven we would all dress, pack up and head to church for midnight mass. This was always harder if the weather was very cold that year. Christmas day was spent visiting with relatives and neighbours. There was a never ending supply of a variety of delicious foods wherever we were at. Nuts, candy and Chinese oranges. Santa would bring gifts but always in form of clothing items as my parents were not great believers in toys. Because of my northern posting during the past three years I had some difficulty purchasing presents for my family. My sister Ollie was now away from home and working in Edmonton, and she graciously agreed to fulfill those duties. My mom always felt I was in need of a new shirt and slacks. Last year she had the local tailor sew a set for me. It was something very special and I still have this shirt more than fifty years later.

During the last week before Christmas everyone pitched in to make things as interesting and as festive as possible at the base. There was a Christmas tree decorated and set up in the lounge area. The RCAF headquarters out of Whitehorse were very considerate with respect of trying to please their staff in this isolated location. Being a male dominated camp it often created some problems in staff relations. A Christmas party was planned and arranged to boost morale and was attended by everyone who was available. I was working the evening shift, but I was able to be in for some of the activities. The three wives and children of the RCAF staff were there. One of the fellows took on duties as a Santa for the benefit of the children. The cook had worked overtime preparing an outstanding banquet. Throughout the lounge and canteen area bowls of confectionary items and fruit were placed for everyone to enjoy. The evening ended with a movie using our camp projector. We had received movies every month or so which were transferred around between the various airports in this region. The movies were nothing special being of the Cassidy types, but it was still fairly good entertainment. During this very special evening we had the option of enjoying some free beer and something stronger being handed out by the RCAF staff in the canteen. It was another Christmas to remember in an isolated, northern posting.

Smith River

Smith River Aeradio building

aeradio control center, Smith River

Airport road washout

Joe Boudreau

Smith River staff

O.I.C. at Smith River, Jim Bresden

Smith River compound

Liard Hot Springs

On Christmas morning 1953 a lot of the fellows slept in. I'd come off the midnight shift and made my way through a quiet base to the mess hall for breakfast. All air traffic had come to a standstill overnight and my only duty being the hourly weather reports. Roger, the cook, was on deck to prepare pretty well whatever we would like. Bacon, eggs and hash browns was always a good standby even on Christmas morning. It was a well known tradition in camp that some of the fellows would make a trip to the Liard Hot springs on Christmas day. These hot springs, named for the nearby Liard River, were first discovered during the building of the Alaska Highway. It was about a fifty mile trip from camp. Howard, one other the power house operators had an old beater of a car we would drive. The RCAF fellows would take their one ton unit. Joe and I planned to catch a ride with Howard, while three of the RCAF boys piled into their own truck. We would drive in convoy- no one would want to be stranded on this road at this time of the year. The temperature on Christmas day was around -25. Howard told me the hot spring area was very primitive, with only an old shack used as a change room. He told me to take along a change of warm clothing as well as towels, blankets and sleeping bags to keep warm after getting out of the hot water. After packing a lunch we departed at 10 am with the cook instructing us that we better be back by 6 pm for turkey supper. The narrow road out of camp was well cleared off with some icy spots. A great deal of care had to be taken around those hair pin turns, and there were some concerns about our driver Howard's eye sight. He wore very thick coke bottle type glasses. Howard maintained a slow and easy pace. Once we reached the Alaska Highway the road condition improved dramatically so we were able to pick up speed. Just before the Liard River there was a primitive sign which pointed to the left. We had arrived at the hot spring site. After parking we walked close to half a mile along a well beaten path in the snow. From a distance we could clearly see the hot springs with the hoarfrost and steam raising high into the sky. Walking out of the snow and cold we entered into some sort of a paradise. All around these rocky pools hot steam rose high above into the surrounding frost covered trees. Near the water there was warmth free of any snow. We piled our gear in a dry spot near the pool and entered the water carefully, knowing some areas would be lukewarm ranging to extremely hot, near boiling. It was a great treat to be able to lounge around in this very unusual outdoor spa. We all rolled around lazily, going in and out of the water, taking smoke breaks and enjoying the hot coffee and sandwiches. As it was Christmas day, there was no one else there other then the six of us from the base. After a couple of hours we felt it was time to head back to the camp. We made sure we dried off properly and let our body adjust to a reasonably normal temperature. With dry and warm clothing back on and with our gear packed we slowly, somewhat reluctantly moved back to our vehicles. The temperature change from the hot spring site to just a short distance back on the trail seemed extreme. The adjustment was pretty well complete by the time we reached our vehicles. Our trip back home was a very relaxed one. I in particular felt completely zapped out being I had come off a night shift and had not yet slept. I managed to have a short cat nap before arriving back at camp. We were back home well more than an hour before supper so I went into my bedroom to open my Christmas present I had received from home.

It was such a pleasure to open the parcel which had a Christmas card and short letter from my parents. As well there were a couple of pairs of socks, a shirt and sweater. Neatly boxed was some of Mom's Christmas cake and cookies. To complete the parcel, two metal cartons of fifties cigarettes. I sincerely appreciated these thoughtful gifts and messages from home.

In the military tradition the table was graced with fresh pressed white tablecloths. Brass candles and a small display of Christmas ornaments were also part of the décor. The turkey was served accompanied by a large variety of other dishes. This was all followed

Embarras Portage: Before, and After

by a Christmas pudding and numerous other pastry dishes. There was no question in anyone's mind that this was the ultimate in dining service. After a fantastic meal I settled in for a well deserved nap and slept like a stone.

Overdue Trappers

Although we hadn't realized at first, it was a week after Christmas and the two trappers who were supposed to spend Christmas with us had not showed up. Everyone knew that for the past number of years they would always arrive with their dog teams a few days before the beginning of the festive season. The sergeant in charge was concerned that they may have encountered some kind of misfortune. As these two men had no known families, the men at the base had kind of adopted them. It was common knowledge that both men endured a very primitive and harsh lifestyle the rest of the year. The staff enjoyed providing them a bit of luxury living during this very special time of the year. After waiting another day with no sign of them, the airbase commander directed a message to RCAF headquarters in Whitehorse outlining the circumstance of these overdue men. Headquarters responded that an Otter aircraft from their search and rescue unit would be dispatched next morning to conduct a search. The Otter would be landing at our site the next day to pick up any spotters that wished to accompany them to the search area. The flight captain would be expecting a briefing of the exact area he would be required to cover including the location of their main cabin near Tobbally Lake and other cabins they may frequent in the South Nahanni region. The weather was very cold, hanging around the 25 degree F area, but otherwise the skies were clear with only a low wind. Conditions were very favorable for the proposed air search. I knew I would be available to go on this trip, and early the next morning we received a flight plan that the Otter was on its way. There were five of us going on the search for the missing trappers so we put on our warm gear and sat around the tower waiting. Around eleven we heard the roar of the aircraft in the vicinity. The pilot radioed the staff on duty that he was making a pass over the runway

Embarras Portage: Before, and After

and requested the wind speed and direction and the altimeter setting. The twin motor yellow Otter landed, approached the tarmac and cut off his engines. The airport manager and pilot exchanged greetings and search instructions were discussed. In a matter of minutes we boarded the Otter for the flight to the search area. All of us "spotters" took window seats in this roomy aircraft and were given a briefing as to what we were supposed to look for. The pilot indicated he would be flying as low as he was permitted considering the rugged mountainous terrain. Soon we were airborne, flying northeast now in the Yukon towards Tobbally Lake. Looking downward it was very obvious the difficulty of searching this unforgiving landscape. All we saw were covered mountain peaks with massive growth of a variety of tress that was followed by deep ravines and sharp cliffs. The pilot manoeuvred the aircraft expertly over it all. It seemed he was very close to mountain sides at times. As we flew I wondered how in the world anyone lived here. We flew over small, frozen lakes and flat areas. The pilot addressed us through the intercom system letting us know that we would soon be approaching Tobbally Lake and to keep our eyes peeled for any sign of life. Making a pass over the sizable mountain lake, a log cabin was noted very near the shoreline at the north end. He made a second pass flying very low right over this site. We all noticed smoke rising from the chimney. This was followed yet by another low pass and that's when we noticed the two men standing outside near their cabin waving vigorously at our plane. This was a clear indication that both men were okay which satisfied the pilot the search was over. The aircraft was put into a climb heading back to the airport. In a bit over an hour this search was concluded and we were back home, our minds at ease. Prior to our departure the cook was advised by the Sergeant to be prepared to have a meal ready for the whole crew. We were all grateful to have company at the base, and to have had a very interesting morning to spend with them with a positive result for everyone involved.

Trappers Arrive

Five days after the search was conducted, both trappers arrived at our base. It was much to their surprise that anyone had been out looking for them. They claimed they had lost track of time by not properly crossing out days on their calendar. They had no other way of knowing the exact time of the month and had estimated it to be only about mid December. These two fellows looked like they had landed from another planet. Both had full beards and long hair giving them a very unkempt appearance. Their toboggans were fully loaded with some furs and personal gear, but mainly with the lake trout which was piled in layers like cord wood. This fish was given to people in married quarters and for our mess kitchen. Some was necessary to feed their dogs too. I was introduced to Oscar and Helmet being a newcomer, and they already knew pretty well everyone else. After tying and settling their dogs in they were free to enjoy visiting with everyone on their yearly annual "holiday". There were some small bedrooms available, so both Oscar and Helmet were assigned places of their own. The first thing on their agenda was to have hot baths and change into some clean clothing. It was now early January and very cold weather had been predicted and that was certainly the case. Trapping is normally at a standstill during the most brutally cold portions of the year, so Oscar and Helmut would be spending a couple of weeks with us. Helmet would be catching a ride to Coal River with our mail run and from there take a bus to Fort Nelson to sell some furs at the Hudson Bay and purchase badly needed supplies. These men were using our base as their address so there was a bunch of piled up mail for them to go through. One of the things they had done was to send ore samples to Ottawa for analysis so they had some replies regarding these requests. Oscar was the main correspondent for this process, and had some additional samples to send out. I was interested in looking at these samples and they appeared very interesting because of the color variations and weights of these rocks. Oscar told me his reply on the samples he had previously sent out were very encouraging. One had six different types of ore but mostly silver. Another ore sample was nearly solid copper. They had their own types of maps of the locations of these ore finds. One of his major concerns was the great difficulty of accessing these extremely remote areas. Both Helmet and Oscar had high hopes of a major company buying their claims and becoming rich. They also felt however, that time had been slipping by and their dream may not be fully realized. It was very pleasurable to sit around and listen to the tales of these two men. Both were well in their fifties and had lived in this region for too many years to count. Helmet was a wiry man over six feet tall. Partially covered by his beard there was a large scar across his face. That, he said, was the result of being attacked by a grizzly. After a great struggle the fight was ended by Helmet splitting the grizzly's head with a hatchet. Oscar on the other hand was a short stocky man about five feet six inches tall. He had a visibly deformed eye due to a run in with a tree branch. Their physical appearance showed signs of wear, and it was obviously due to a very harsh lifestyle. It was also quite clear that both these fellows were in top physical shape. We all found it very difficult to understand that after living in total isolation for long periods of time they displayed very likeable personalities and each had a great sense of humour. While Oscar helped the cook around the kitchen and with cleaning chores, Helmet caught the mail run to Coal River and from there headed to Fort Nelson by bus. On the way back he would arrive at Coal River on the date of the mail run to catch a ride back to our base. He took with him a small number of furs to purchase some supplies for their current needs on the trap line. Helmet would arrange with a bush plane charter operator to fly in to Tobbally Lake after spring break up about mid May. The pilot would bring a plane load on a Norseman of all their basic needs to last them through the summer months. On the return trip the float plan operator would take out the winter trapping catch, some to be left at the Hudson Bay store but a majority being shipped out to a fur auction in Winnipeg. The charter trip to this very remote lake was very expensive. For example it cost about two hundred dollars to bring in a hundred pounds of flour. Oscar claimed that their winter trapping was usually very successful and fairly profitable. There was a great variety of animals they trapped: martin, mink, lynx, fox, wolf, coyote and other small species like squirrels and weasels. But it wasn't about the money. Oscar talked freely about their years in the Nahanni and their love for the land few people journey to. All summer long they would each wander off by themselves climbing and struggling over the difficult terrain looking for precious ore. They had learned to be extremely careful in every step as help was not existent. Every potential find would bring about a rush of great excitement. With an ore pick in one hand and the Winchester in the other was a way of travel. Running into a grizzly was a frequent occurrence. One particular find stood out very clearly. Many of us gathered around, listening with rapt attention as Helmut described the day that with the summer sun high above him he came across a whole mountain side of high grade copper, glowing like a wild dream.

The Duck Hunt

During the coldest days, with nothing better to do my mind would snap back to life at home on the farm, to when I was about fifteen years old. With harvesting on the way, duck hunting was also a pastime. The birds were going south but feeding on the grain still in the fields. Usually on Sunday afternoon a few of us fellows would get together to see what we could bag. There was Johnny Gustha, a friend my age who lived one half mile east of our farm. Peter Paziuk was another fellow who resided two miles to the west. None of us possessed any great fire power or have funds to purchase ammunition. I had a sixteen gauge shot gun of my grandfathers. It was an old beater that seen better days. Johnny had a twelve gauge double barrel that was in not much better shape. Peter was the worst off having only a single shot .22, however Peter was lucky enough to have a box of .22 shells. Shot gun shells were a problem so we had to make our own. We would first go out and gather shells left on the fields fired off by the rich grain elevator agents from town. There was many to find, both .16 and .12 gauge. We would collect the casings and take them back to our blacksmith shop where shell restoration would take place. First step was to knock out the cap with a large flat nail. The bent in portion was hammered out with a shingle nail. Match heads would be carefully inserted with the claw adjusted, then carefully again hammering it back into the shell casing. When this was completed, shell loading was the next step. My dad had purchased a metal container of black powder a couple of years earlier. One of the things we did not have was lead buck shot so we had to be very ingenious. We knew the five gallon milk slim line cans had stoppers that were made of solid lead. There was usually one around the farm that was no longer serviceable. By removing this stopper it was placed on an anvil and hammered out into a flat sheet. With a hammer and chisel we cut out small chunks of lead to be used as suitable buck shot. We now had everything necessary for loading. First we poured a small supply of powder into shell that was packed down with newspaper than pour in some buckshot, and pack it down with more newspaper. This process was repeated a number of times until we felt we had sufficient shells for the hunt. The power of the shells was solely dependent on the amount of powder we used. The best spot for duck hunting was near a small lake where the ducks would fly off the fields after feeding, enroute to the watering holes. With stooks we would build blinds and wait for the ducks who would be flying very low before entering the water. The best time for the fire power to commence was just before dusk as that's when the flying in would begin. The blinds were built in a circular fashion and large enough to accommodate all three of us. We laid down stooks inside to sit on opposite from each other. Then we smoked away our cigarettes and waited. The first sign of darkness would slowly begin to close in and as suspected there would be small flocks starting to land on the lake. These would act as decoys for large groups to come in. From our fort we made preparations by laying our ammunition for easy access. Then without much warning all hell seemed to break loose. Flock after flock approached the lake making descent over our blind. All of us quickly started firing at these low flying birds. Because of the closeness and number of ducks in these flocks some of them started to rain down, some dropping dead and others only wounded. An odd shell would misfire due to the imperfection in our reloading. Black smoke and flying newspaper filled the air and we continued our offense until our ammunition ran out. Peter still hammered away with his 22. We gathered up our rewards which amounted to a couple of dozen mallards. We equally shared the bag and headed home carrying our load. Mom would be pleased with our accomplishment and there was roast duck on the table for the next couple of days.

Winter Blues

In January, 1954 the winter conditions in Smith River moved in with a vengeance. There were blizzards, with heavy snow fall and strong winds followed by clear skies, lighter winds and extremely cold temperatures; in the forty to fifty degree below zero range. The conditions seemed to carry on for weeks without any reprieve. We were mostly confined to quarters which could be depressing. The two trappers were getting kind of anxious to return home and to their trap lines, as much as they enjoyed and were appreciative of the comforts of our airbase. Our evenings were spent in the canteen lounge. The weekends always led to poker games that lasted into all hours of the night. I was mostly a watcher. I did not see any sense in wasting my hard earned money in this fashion. The boys around the table gambled and drank beer in great quantities. As the night would go on they also got hungry. Sometimes they would ask me or some other of the fellows to go to the kitchen to make sandwiches. We didn't mind. In fact, this would prove to be very profitable as I received substantial tips for the service. Usually the bigger the winner, the larger a tip we received. Those of us not playing became very proficient with our sandwich-making and catering services.

During the harsh weather conditions a large amount of time would be spent in and around our aeradio tower centre. Sometimes simply playing a game of crib or having a good visit with the operator on duty. Jim, our young officer in charge, was always available to bounce off any problems I might be encountering. Jim seemed to like my work habits and we hit it off real well. He strongly suggested that I improve my standing as an aeradio operator by taking upgrading courses offered by the department. I was currently at a level one rating, the basic level of all operators in the non supervisory category. Levels two and three positions are eligible for officer in charge appointments. Of course, the two higher levels also meant an increase in wages. Jim was at a level two category but continued to take courses working towards level three rating. I agreed with Jim this would be a good time to improve my standing and work towards a supervisory position in the future. Jim made sure the lessons and text books were forwarded to me asap. These lessons included more detailed study of aeronautical procedures and regulations, such as instrument landing systems. Also information on various equipment we were required to operate, service and maintain. Having all the time in the world when I wasn't at work, I put all my spare time towards these upgrading courses. Jim was helpful with coaching me through any areas where I had encountered some difficulty.

Like the snow which surrounded the base, time drifted gently into the month of February. Weather conditions did improve that month, to some extent. The slightly warmer temperature gave the two trappers an urge to pack up their gear and leave for their home. They were interesting company around the base, so it was kind of sad to see them go. Their dogs were very excited to be hooked up to the toboggan after a slow few weeks. As we waved good bye, they called back they would be back after the end of trapping season around May or early June.

One morning while sitting around the lounge area in our barracks we could smell a very pungent odour. That's when we noticed smoke coming out from under the door of Joe's bedroom. Albert and I quickly rushed over and threw opened the door. It was full of smoke and was difficult to breathe. Joe was out like a light, we got hold of him and carried him straight outside to a porch area. This excitement brought in other staff members rushing in to see how they could help. It wasn't difficult to see what had transpired. Joe was a heavy smoker and after coming off a midnight shift and having breakfast, he retired to his bedroom. After getting into bed he lit a cigarette and promptly fell fast asleep. The cigarette dropped on the feather pillow but only smothered. Joe was given a lot of fresh air, which seemed to revive him, followed by excessive coughing and he quickly came back to life. While this was going on, Albert and I carried out his smoky bedding which included the partially burnt pillow. These items were deposited in the snow outside and soon after caused the pillow to break up in flame. After Joe recovered, The RCAF sergeant in charge was not very impressed and suggested in no uncertain terms that Joe control his smoking habit while in his bed. Despite Joe downplaying the severity of the situation, it was very obvious that he would have suffocated if Albert and I had not been there to intervene.

I applied myself with studies throughout the winter working toward my Level 2 Aeradio upgrading as Jim had suggested. I would also write letters home frequently as this was my only way of communicating with them. I was always pleased to hear from my family. My sister Ollie who lived in Edmonton would write me. Occasionally Dad would drop a few lines to let me know how things were with them. One of the operators, Doug McKay, on annual leave to Vancouver returned back to work in February. Bill Andref, a married operator on temporary duty at our station got a transfer to Medicine Hat so he could be reunited with his family. This meant we were now maintaining a full staff of four operators and an officer in charge. Joe liked to take his month of annual leave in the spring. He usually only went as far as Edmonton and after spending a month living a carefree lifestyle he would return back broke and exhausted. Joe would never come clean as to how he had spent his vacation but we had a good idea of the nature of his activities. During periods of annual vacation if a replacement operator was not available it meant working overtime shifts. No one really complained about this being the extra money was always a welcome boost to the bank account. I had difficulty at that age understanding some of the people that lived and worked these isolated locations. Some of us had a purpose by coming here, with an aim of saving money and a view of a better lifestyle in the future. There were many others that landed in places like this with no future goals whatsoever. A considerable number of these fellows were pushing middle age, saddled with a history of marriage breakdowns and other problems like addictions to alcohol and gambling. Some blamed their years of overseas military service for their uncertain future. One of the fellows I clearly remember was named Clifford Hall, a tall mustached middle aged ex-military type. He had been posted at Smith River just over a year when I was there. He was separated from his family and worked as a power house operator. Clifford drank and played poker to the extent he was usually broke from pay-day to pay-day. His annual leave was due, and he had plans to go to Edmonton by bus. The trouble was he was short on cash and approached me about a loan. He had a brand new 30-30 Winchester rifle as a security for a hundred dollar loan. Feeling sorry for him I agreed to this transaction and I still have this gun today.

Visit by Game Department

March, 1954 started to show some signs of spring oncoming. There were more hours of day light and on an odd day some melting did occur. One sunny day while Doug was on shift he noticed a flight plan for a Cessna 185 with destination being our airport. The aircraft belonged to the Yukon game department out of Whitehorse. We remembered hearing something about having a wolf problem just north of us in the Nohanni. Apparently the wolf population had increased exponentially and they were killing the moose at an unhealthy level. The ETA for the game department plane was around noon. A few of us went to the tower to await the aircrafts arrival. As the flight approached our airbase, the pilot radioed in requesting landing instructions.

Smith River Radio this is Cessna CF-NOP ten miles south starting descent VFR

CF-NOP – this is Smith River Radio" Roger – check position – altimeter setting 2990 – wind 280 degrees at 5 mph no reported local traffic – land at your own discretion – ski runway snow rolled and is compact.

Smith River Radio this is CF-NOP Roger – downwind for runway 29.

CF-NOP this is Smith River Radio – Roger – CF-NOP down 38 – Roger – Air traffic control in Edmonton and Whitehorse Aeradio advised of arrival.

The ski equipped aircraft taxied in and was greeted by the airport manager. There were two on board- the pilot and game officer. We left the tower to meet these two fellows. The manager invited them to join us for dinner. As usual it was a highly rated feast. The manager always made a point of inviting any visitors to take in a meal offer. As we ate, the game officer outlined the reason for their trip. They had poison bait on board which would be dropped off at a couple of lakes north of here. One of the lakes was fairly large, called Dead Man's Lake located fifty miles straight north in the south Nohanni region. A smaller lake nearby was second drop off point. The lakes are used as a safety factor for other animals. The wolves often cross lakes in search of food, and likely they would locate the bait easily. Then once the lakes melt in the spring the uneaten bait would sink to the bottom of the lake, preventing other animals from ingesting it. The two men did not wish to hang around the kitchen table too long, wanting to get back to Whitehorse before dark. Within fifteen minutes they were back in their plane and the motor roared to life. The pilot radioed the tower for take-off instructions and advised Doug they would be returning after the drop off in approximately one hour to take on some extra fuel. The plane was off in a matter of minutes. As we had expected the Cessna radioed in some 50 minutes later for landing instructions. The pilot had previously arranged with the RCAF manager for about thirty gallons of aviation fuel. All the aviation fuel used mainly for emergency purposes was stored in 45 gallon drums. There were about one hundred drums laid out neatly near the tarmac. As the aircraft taxied close to the storage area one of the RCAF fellows went out to assist with the refueling which was carried out quickly with the aircraft starting up its motor and taxing towards the ski runway. The pilot radioed in, filing a flight plan back to Whitehorse. Shortly after refueling he was airborne and Doug provided him with enroute weather conditions. Their flight time back to Whitehorse would be about three hours, which would get them back home before dark.

    I was curious about the storage and emergency use of the aviation fuel we kept on site. One of the RCAF maintenance fellows filled me in. Every year a fresh new supply of this higher octane fuel is brought in by trucks. It always came in 45 gallon drums. When required it could be provided to Government Aircraft or forestry services. It could also be given to aircraft that for one reason or another are running low on fuel and could not reach the next closest airport. After a year the remaining fuel, which is very valuable, is condemned and put up for bids. Very few people in this neck of the woods would have any use for condemned aviation fuel except one. That was Tom, operator of the Coal River Service Station. He would put in a one dollar bid on a 45 gallon drum of fuel. Tom was lucky in securing sometimes up to fifty drums at that price. At his service station he would mix aviation fuel with diesel oil to reduce the octane level, making it operational for vehicle use. He had two pumps in use at his station, one for providing local people with proper gas, and the other for selling the mixed fuel to travelers along the highway. He was able to pocket a great profit from this ingenious, but unscrupulous operation.

Whitehorse Trip

Spring was in the air and everybody was trying to shake off the winter blues. Our mail truck driver reported the Alaska Highway was in fairly good shape. Every so often people from our base would make plans to journey to Whitehorse, for no other reason than a change of scenery. A trip could be made only before the spring break up occurred; afterward the road to the highway became extremely difficult, mainly due to the mountain run-offs, which caused wash outs in a number of the places along the route. I was very interested in taking this trip if I could work out the necessary shift changes. Bernie McDonald, an RCAF Corporal, planned to go with his wife and two small children. They would have their own vehicle. Howard Jackson had an older model car, and said he would also go if he got enough passengers to help out for the trip. I discussed it with Jim and he agreed I could leave if a suitable cover off for shifts could be arranged. William, the weatherman, wanted to come and would also require radio operators to cover off for his day shift observations. The only other interested passenger was Peter Milick, a civilian power house operator. We all agreed to go the following Thursday and return on Sunday giving us two whole days in Whitehorse. Being that Whitehorse was over four hundred miles away, both Thursday and Sunday would be taken up by travel. None of us had any particular reason for going on this trip other than seeing new country. We thought it may be great just to enjoy hanging around this northern city, see a movie and visit a bar or two. Bernie and his family had planned a shopping trip, and he was interested in upgrading his vehicle. We would travel in convoy due to very unpredictable road conditions. Tire chains and survival kits were a must along this isolated highway. All three passengers provided Howard with fifty dollars each for the car expenses. Shift cover off arrangements was made and we were free to leave later in the week. A very early departure was planned. On the day of the trip, we headed out well before dawn. The road to the highway was a nightmare as usual. Once we reached the highway conditions improved. The highway itself was free of snow. We picked up speed, our first stop at Coal River where we all had coffee and gassed up our vehicles. We agreed on stopping every couple of hours along the way to stretch our legs. As we drove along this barren landscape, I wondered why places such as this existed. There were the steep hills to climb and deep ravines to contend with along this narrow and twisted highway. Occasionally we would meet a transport truck and Howard had to be extremely careful in passing as any kind of an accident in this neck of the woods would be a disaster. Our next stop was a roadside café and service station at Watson Lake. After that was Teslin, a much more scenic location with a long lake we had followed for many miles. After a late lunch and checking out our vehicles we continued to Johnson's Crossing. This was our last stop before arriving in Whitehorse nine hours later. The last few miles before arriving in the city seemed drab this time of the year, but it was obvious the area surrounding Whitehorse with its rivers and mountains would be a very beautiful place during the summer. The drivers suggested that the "Klondike Inn" would be a suitable place to stay. We checked into this very northern type of accommodation. William and I shared a room. For the next couple of days we enjoyed lounging around this spectacular wilderness city. Attractions included a variety of restaurants, bars and movie theatres. The city itself was nestled in the banks of the famed Yukon River, and surrounded by mountains and clear mountain lakes. We could only imagine what this place looked like during the spring and summer months. After tiring ourselves out with sight-seeing and entertainment, we were all quite prepared to head for home on Sunday. Our bunch did little shopping but Bernie had traded off his car for a brand new Chevy. As planned our departure was on schedule leaving Whitehorse at day break on Sunday. We would be stopping at Johnson's Crossing for breakfast. The weather, though fairly warm, was very cloudy and there was an indication of rain ahead. We followed Bernie but as he was a bit heavy on the accelerator, it was difficult to keep up. Perhaps he was trying to show off with his new car. The road became monotonous at times causing some of us passengers to drift off to sleep. It wasn't until we were somewhere between Teslin and Watson Lake that we were starting to encounter some difficulty. We had run into freezing rain, which began causing problems climbing some of the steeper hills. It wasn't long before we ran into a Freightways truck chaining up to climb an upward slop of a high hill. Both of our vehicles continued for awhile up the high grades with a lot of spinning and sliding, but as the freezing rain continued it we realized it wouldn't be too much further before we would have to put on chains. We passed at least one vehicle which had slide off the road, but did not stop as there was no way we could help to get it back on track. As predicted, our luck ran out on the next long incline. Harold decided it was time to chain up, but Bernie was a bit disappointed as he felt the chains banging around the inside fenders could cause some damage to his brand new car. It was a very messy job but we all got to help, rolling around on the wet, muddy ground. Once the chains were secured we were off at now a much slower pace. The loose chains caused the ends to bang within the fenders in a monotonous rhythm. Harold, our driver, was pleased as he was now able to climb the steep hills with ease. Despite the constant banging some of us dozed. After fifty miles of freezing rain we were in the clear. We soon arrived in Watson Lake, where we stopped at a local roadside café to clean up and enjoy a long overdue lunch. The trip from Watson back to our base went smoothly. Darkness was beginning to settle in by the time we arrived at home. Supper hour was over but we managed to drum up a few leftovers. I was back on day shift with the following morning so I retired to my room early.

In the weeks that followed our trip, it was clear that spring had returned. Snow was melting with great speed, and streams of water runoff were everywhere in this mountainous region. Looking at myself one morning I noticed a distinct pot belly, due to the extra weight I had put on over the winter months. This was a common scene on this base, no doubt caused by the excessive intake of exceptional meals and the sedentary life style. Albert had a similar problem so he and

Embarras Portage: Before, and After

I talked about getting an exercise program started. We decided on running, and with the snow nearly gone off the runway that would be a suitable track. As with any new undertaking, at first it was a most painful ordeal. As time went on we were both running the mile long runway with much greater ease. That, along with greater care taken at meal time, and we soon noticed the change in our physical shape. At my age, I would have been ashamed to go home in that condition. Speaking of home, I had received a letter from Dad who stated both he and Mom were preparing for spring work. They were so pleased about the fact that Stan and I had purchased the two quarters of land from Mr. Chimko. No longer will they be required to pull their farm machinery back and forth to St. Brides. With the over two hundred acres of land nearby was a dream come true. As I read this letter, thinking about the hard work my parents were getting ready to undertake, my mind drifted back to one summer job I'd held that was so distasteful that even now my children could never understand.

Summer Job

At the completion of my grade twelve and being accepted into Radio College I needed a summer job in Edmonton. My friend Johnny Gustha and I went into the city to start our search. Johnny found an ad in the Edmonton Journal that ten people were required by Swifts packing plant. Both of us applied and without any difficulty were accepted. The job was in the warehouse section of the plant. Before starting, the foreman explained that this work we were about to do would last about ten days. If we were able to complete it, we would be assigned to work jobs in the main plant. As we entered the warehouse it was all too clear to us what was in store. The smell was that of curing cow hides. There were thousands of hides all folded and piled one on top of the other. The skinned areas of these pelts were salted down and cured for a number of days. It was now time to remove the salt from the pelts and send them in for tanning. The smell was similar to that of decaying animals. To remove the salt, the hide was laid out and then banged on a wooden ramp. Six people would hold on, three on each side of this hide, one on each end and one in the middle. We all took turns yelling out instructions, up-down, banging the hide against the ramp. The banging would cause the salt and other juices to fly off in all different directions. Soon we were all covered with very smelly juices and salt. There were nine persons assigned to each operation. One to throw down the hides, six on the ramp duty and one each for refolding of the hides and the last one to re-pile. When we entered the cafeteria in the main plant during the dinner hour we soon notices everybody moving away from the area we sat down for lunch. I guess we kind of got use to the obnoxious smell not realizing how it would affected other people. We talked to the grumpy foreman about this and he did provide us with some well used smocks, which was not a total answer to the problem, but did help. Riding the bus to my aunt Sophie's home where I stayed was not too fun either, with people moving away from wherever I sat down. Everyday there were new fellows hired, but also every day some left, some only trying out the ramp for an hour or so. My friend Johnny also left after being there a couple of days. Originally there was to be two full ramps serviced but there was difficulty at times even to maintain a full staff for one. The job foreman was not a very nice man, which didn't do much for the staff morale. He would go around barking orders and limiting our smoke breaks. The salty, slimy hide juices began the raise heck with our hands as well. As the days went by I kind of got used to the routine and the very unpleasant work conditions we had to endure. One of the other staff was a young fellow like me, who said he was attending the University of Alberta and needed this work. There were so many changes in the other staff so it was hard to keep track of any of them. Every day we watched the pile of hides get smaller. After twelve days, our job was finally completed and there was a great cheer from the crew when the last hide was hammered out. The few remaining fellows, including me, who completed this project were given slips by the foreman to report to the main office the following day. We would be assigned jobs in the main plant. It was a huge relief to get out of those smelly clothes. My good aunt would wash them for me every night, for which I was very thankful. I was very excited on getting on the bus and going to work the following day. I wondered what new assignment at the plant would be. In the plant office I was greeted by a supervisor who accepted my slip. He sat there, then reviewed a bunch of records and informed me that my new job would be on the killing floor. I am sure he noticed the expression of disappointment on my face. It was then that I told him I could not do that type of work. I had only one experience in killing an animal and it was extremely distasteful to me. A few years earlier, at home on the farm I had to butcher a hog while my dad was working in a bush camp and my mom found we were out of meat. My mom said I had to do it, as even my older brother Stan could not handle that. It was a very unpleasant memory, which I won't go into detail about here. After outlining this to the supervisor he must have felt sorry for me. He again reviewed the records for openings in other departments, and was able to come up with a job in the sausage department. This type of job appealed to me much more and I thanked him for his consideration in my case. He told me that anyone who could stand and complete a job in the hide department deserved a good placement.

Working in the sausage department proved to be a rewarding experience. After being greeted by the supervisor, I was given a clothing issue of white cotton shirts, slacks, hat, gum boots and a smock. My place of work would be on the second floor. My main duty was to load six compressor tank units with the raw mixture of sausage, wiener and baloney meat. I would load this material into a stainless steel wheelbarrow and use a stainless steel shovel to load the compressors. Each compressor was operated by a young lady who fed the raw meat into appropriate casings. When the compressors were properly serviced my second duty was to continuously pressure water hose down the whole working area of any meat spillage. The concrete floor was always wet, thus the need for gum boots. Health inspectors frequently showed up, so cleanliness was important. Other than working in very damp conditions I enjoyed the job a great deal. I got to know a couple of the young girls operating the meat compressors, which made the job more interesting. Once the particular product had the casing filled it was hung on metal racks which would then be wheeled into cookers and or smokers for final preparation. We were allowed to sample any of the completed product which was often our lunch. About a week before going to radio college I quit this job with mixed feelings. I got to know many nice people at my first real summer job, so I was a bit sad to leave.

Springtime in Smith River

Spring was upon us in Smith River, and it was a lovely change even in this neck of the woods. I continued to work hard on my studies for the level two rating. Jim was my mentor and forever applying pressure on me to study, and although he was tough he was also very helpful and motivating. It was obvious that Jim liked my style of work, and we seemed to share similar goals in life. We all knew that Jim had spent his last two annual leaves in Edmonton taking flying lessons. He mentioned to me that he had 100 or more hours of flying time and was working towards getting his commercial rating in the future. He was planning to go out on annual leave in the spring, and this year was hopeful to buy a small plane. In the meantime he was also studying for his third level examination with a view of entering the technical field in the area of instrument landing systems. His flying experience no doubt would have positive bearing on his ambitious future plans. As for the rest of us, Joe Boudreau at least preferred to take his annual leave in the spring so approval was granted for May 15 to around June 15. Jim applied for his leave to start once Joe returned. We had received word from our regional office to the fact that there was a shortage of replacement operators, so overtime would be necessary. This news was no great disappointment to us. The extra money we would receive was always welcomed. Alan and I continued to hit the runway with our exercise program. It was clear that we were succeeding to some degree. Alan, however, was encountering some difficulties with his diabetes condition. He did not take great care of himself and often suffered blackouts. We were all aware of this and would come to his aid whenever it happened. Being a war veteran, Alan had the privilege of attending a veteran's hospital in Edmonton to stabilize his condition. He went every spring for two weeks so he planned for his leave at the end of May. When he came back we did notice a change in his attitude, but soon fell back into the same old routine. When Joe left on his annual leave, he mentioned going back east to see his family but when he returned in mid June he told us he got only as far as Edmonton and hung out with some of his old friends. Through the grapevine we learned that the friends he mentioned were some undesirable ladies, who left him very broke upon his return to the base. Upon Joe's return Jim proceeded with his holidays. Joe was the senior man on staff, but despite his seniority was not that reliable. Jim assigned Gil to act in charge during his absence. There was very good cooperation amongst our staff so there were not major difficulties to contend with. Everybody was working overtime shifts and did not have much time for anything else than that.

Two weeks into Jim's holidays our morning shift operator noticed a flight plan with the destination being our airport. It was a Cessna 140, identification CF-FCR. To his surprise he also noted the pilots name being Bresden. Without a doubt this was our boss coming back two weeks early, with his own plane! The flight plan indicated he had an overnight stopover planned in Fort Nelson where Jim had a close friend stationed. Early the next day his flight plan was updated. The destination was our airport. He would be flying VFR via the Alaska Highway, ETA 1:30 pm. Word got around the base quickly of this unusual occurrence so we were all waiting patiently for Jim's arrival. I was working the afternoon shift. We hung around the tower awaiting radio contact with Jim. Sure enough, at approximately one o'clock a radio signal addressing our station was received.

Smith River Radio this is CF-FCR –

Doug responded *CF-FCR this is Smith River Radio Roger go ahead.*

Smith river radio this is CF-FCR – Roger my ETA (estimate time of arrival) is 45 – cruising 5000 VFR via Alaska Highway near turn off

CF-FCR this is Smith river radio Roger altimeter setting 2997 – wind 290 at 10 mph – no reported local traffic. Smith river radio – this is CF-FCR roger.

Half hour later radio contact was again engaged – *Smith river radio this is CF-FCR – have runway in sight inbound for runway 29*

– CF-FCR – Roger wind and altimeter unchanged – land at your own discretion.

Smith River Radio this is CF-FCR down 47. CF-FCR Roger –

Doug promptly dispatched the arrival times to both ATC Edmonton and Aeradio Fort Nelson. It seemed half of the airport was at the tarmac to meet Jim and his aircraft. This small, sleek, brightly coloured yellow aircraft landed smoothly, moved into position and cut off the motor. Jim was surprised to be greeted by such a large gathering. We all looked over and admired his aircraft. Jim's first chore was to find a suitable parking place. Afterwards Jim went on to tell us how this purchase had come about. After his arrival in Edmonton, he immediately started looking for a small aircraft. He saw a listing at the flying club that there was one available in Calgary. He promptly headed there and after carefully looking over this specimen and studying its history, he fell in love with it. A deal was made and he flew back to Edmonton. From his work habits here at the airport we all knew how meticulous he was; and we knew he would maintain his new toy in the same way. Jim still had over two weeks of holidays left but he decided to have them carried over. On his free time he would be able to fly around this area. He felt it would be a learning experience flying in this mountainous region. Some of the factors that appealed to him about this plane was its new, higher powered motor and long range tanks that offered a fly time of six hours at a go. We all knew that any spare time Jim would have would be spent around his new airplane. As time went on Jim often took on a passenger to join him on a trip, some being very short flight and others lasting more than an hour or two. As Jim and I had become pretty good buddies I was a frequent passenger. Jim was a great teacher and he would often encourage me to take over the controls. Working with aircraft one gets to know the basic functions and some of its operational characteristics quite well. It was a great deal of fun flying in and around this unforgiving land of the Nahanni region. Landing the aircraft required a great deal of care and attention and is perhaps the most dangerous part of flying. Occasionally when approaching the runway on descent Jim would instruct me to line up. This was always easier said than done. From

3000 feet or so the runway appeared to be about 10 feet wide; not a very easy task without a great deal of practice and skill. I never did accomplish this with my limited flying experience. Any spare time Jim had he spent around his pride and joy. He kept piling hours towards his commercial rating, often commenting on how expensive it was to own and operate his own airplane. The insurance was very high and every one hundred hours of operation it required a complete inspection by a certified aircraft mechanic. Those of us who went flying with Jim did chip in for aviation fuel which I'm sure he appreciated. Without any doubt, Jim was the most conscientious person I had ever met. Before each flight he would meticulously check each part of the aircraft for any possible malfunction.

As spring drifted easily into summer, everybody had their annual leave planned out. I did not make any plans, as I was quite sure I would be getting my transfer during the later part of fall. With the staff shortages overtime was an ongoing occurrence and kept us all busy. If time permitted I would occasionally take a trip with the mail run to Coal River. William, the weatherman, was an avid fisherman. He was introduced to a small mountain lake two miles walking distance from the base. There was an ample supply of rainbow trout which he frequently caught and brought to the mess for our cook. I would sometimes travel this short but difficult trail to the fishing hole with him or another person from the base. We would set up a campfire and prepare an outdoor cook, which was very peaceful and relaxing way to spend a couple of hours and reminded me of what things had been like in Embarras Portage.

Other than these occasional distractions, life around the airbase was routine. There were still the weekend poker games, with the high stakes gambling and excessive amounts of beer drinking which were usually held at the canteen. Bernie, the fellow who purchased a new Chevy on our trip to Whitehorse, would show up at the games saying to the players "look what I was able to buy with *my* winnings." That may have been an exaggeration, but no doubt, Bernie was a very shrewd player and used every opportunity to his advantage. He even lent money to losers, with repayments set on pay days. I still provided a sort of midnight catering service to the players whenever

Embarras Portage: Before, and After

I was available, which continued to be an entertaining and profitable pastime.

River Trip

Albert and I used the runway regularly for exercise that summer. I was able to trim down, lose some of the extra weight that had piled on during the winter months. Fall was just around the corner. The nights were cooler and the landscape became a colourful natural display. It was during one of our runs that Albert asked me if I was interested in a trip that had been presented to him. Bernie, the air force fellow, wanted Albert to shoot a moose for him. If the hunt was successful, he would pay Albert two hundred dollars. Albert wanted me to join him on a river trip to hunt down a moose. He said he had conducted a similar trip last year and was successful. He had shot a moose on a river bank, and with the help of one of the other fellows managed to bring it back to the base. The river trip takes a very unusual course, winding around the airbase in a rough horseshoe. It wasn't an easy task. There are only two approaches to the river. To enter heading downstream, one had to travel along three miles of mountain trail to the river's edge. For about twenty miles, the river flows in a mess of curves and bends, ending about two miles near an exit road back to the base. Albert considered this river ride to be very wild and exciting, but safe as long as the proper precautions were taken. The RCAF owned a sixteen foot flat bottom boat which we would be able to borrow. Due to the swift current, a motor was not necessary. The trip would take place once the hunting season opened. Albert would make all the necessary arrangements, and I agreed to accompany him, being we had become good friends and he needed a helper. Albert assured me that if we were successful in the hunt, he would share some of the $200. For me, it was about the adventure and not the money. Besides, I had a strong feeling that perhaps this trip would be a means of a repayment to Bernie more than anything else. Albert loved to gamble but was often a loser, and Bernie gave him advances regularly which I'm sure Al wasn't always able to pay back. Any money I received on top of that would simply be bonus. Years earlier someone had prepared a map of the Smith River which outlined the route we would be taking. I had a chance to survey it. There was a clear indication of a very rapid current flow off the mountainous areas throughout the whole trip. On top of that, there were also white water rapid areas to contend with. Although it was a little nerve wracking to see it all laid out on paper, this map gave us a rough idea of some of the terrain we would be encountering. It was the middle of September when a date for the river trip was agreed on. Prior to that, Albert was busy making preparations for the excursion. I had my 30-30 Winchester. Albert arranged for a 303 from the RCAF supply security store. We would also require knives, a meat saw and hatchet. Camping gear for coffee breaks and noon meals was on the agenda.

Finally, the date arrived and we were up at day break. After enjoying a quick breakfast our gear was packed up. Albert had put together a packsack with food and camping gear. The weather was on the cool side, but the skies were clear with little wind. Bernie, the RCAF Corporal had the boat loaded on the trailer and ready to be pulled by one of their four wheel drive vehicles. Soon we were all packed up and moving slowly along a narrow mountain trail to the river launch. When we arrived it was all too clear how swift the river current actually was. We both had paddles to steer our way along the river, and our boat seemed to be a very stable, but it was obvious we had to use a great deal of care and attention once we got mobile. At the launch site the river was only about two hundred feet wide, but Albert stated it would vary along the journey. As I had experienced a considerable amount of river travel while in Embarrass, this trip- although different- did not appear too much to handle. Bernie said he would meet us at the end of the trip later that afternoon. He understood our arrival time was dependent solely on the hunt. If we were lucky enough to shoot a moose, it would take time to have it prepared and loaded. He told us not to hurry, and in any event it was okay with him however long we took. After loading everything into the boat, we were off. I was excited for this new adventure. The first leg of the journey was rough, and we swayed along in a rock and roll fashion. This was followed by a part of the river that gave us a smooth ride and again followed by more turbulent areas. After about an hour fighting with our paddles we decided to land on a sandy beach area for a smoke break. We lit a fire on the sand in this peaceful nook and brewed some coffee. Al noticed animal tracks, some which we felt were that of a moose, which was a positive sign. Soon we were off again, and as we cruised along we kept our eyes peeled for the purpose of our trip. We would shoot a moose only if it was on the river's edge, since packing the meat any distance would be out of the question. We were heading in a northerly direction- now in the Yukon's south Nohanni region. As we sailed along we spotted a large grizzly on the water's edge, sending uncontrollable chills up our spines. He did not make any great effort to move as we passed him- no doubt he was curious of our presence. We decided to make fairly frequent stops if and when we came across suitable landing spots, the purpose being mainly to view the area for possible drinking spots for moose. Yes, this was pretty wild country and it wasn't much longer before we noticed two timber wolves gracefully moving along on the bank of the river. They displayed very little interest in as we cruised past. Noon was fast approaching so we decided to look for a landing spot as the work of paddling had made us hungry. The spot we choose made an excellent camping spot, a sandy area with a good supply of dry wood nearby. We built a fire on the sand near the water. The meal was enjoyable, with the main course being pork and beans with fried pork sausages followed by a couple cups of tea and a smoke break. After a cleanup and repacking our gear we once again shoved off downstream, with no further sightings of wildlife. We were proceeding in an easterly direction then. The river seemed to be narrowing and piled up by drift wood. In a few locations we had to be very careful in passing some jagged timbers which could have caused damage to our boat. As we came around in a bend in the river we were surprised by a bull moose enjoying a drink of water. This caused an immediate flow of adrenalin as we headed to shore trying to get into position for the kill. The commotion caused by this activity seemed to make this large animal leery, and it made a turnabout and headed towards the deep woods. By the time we parked, Mr. Moose was out of sight. This was very disappointing, mainly for Albert's sake. We managed to get out of the boat with our weapons and surveyed the general area hoping our target was still nearby. With no such luck we were back in the boat now heading south. Once again we entered an area with rough, turbulent waters which bounced us around but luckily eddies kept our boat on a straight path. After passing through this critical area we once again had fairly smooth sailing. At one point Albert stated to me he seemed to be encountering vision problems. He suggested we make a landing at the first suitable spot. We pulled into the nearest bank and I immediately tied the boat to a tree. Albert knew what the problem was. He was a diabetic, and most likely requiring a sugar intake. He had a supply on hand and after reaching into his pocket he ingested a couple good belts followed by a drink of water. This was not the best spot but we decided to fire up and brew some coffee. This break seemed to stabilize Albert and we soon moved on again on our mission. Prior to leaving for this trip I was somewhat apprehensive knowing Albert's medical condition. It had occurred to me what action would be necessary in the event he suffered a blackout. Being out in the wilderness riding a boat that required a great deal of attention for both of us, we had to be ready for all possibilities. Luckily, Albert showed signs of full recovery and there were no further incidences relating to his condition that day. We hit an area of water that gave us a fairly smooth ride so we were able to relax a little more. This river was the most unpredictable I had ever encountered. It seemed to vary in width from some fifty feet to about two hundred feet. We could enjoy a smooth ride which would quickly turn to a very turbulent and unstable white water section. As we cruised along Albert spotted another bull moose some two hundred yards away, just standing there on the river bank. This was the second chance we needed. We slowly manoeuvred our boat to the opposite bank. As I jumped out to have it stabilized on shore Albert grabbed the 303 and with a great deal of excitement took aim and fired at his target. The recoil rocked the boat and the large animal belted a loud roar then stumbled and fell. We quickly redirected the boat towards the fallen moose however by the time we reached him he was up and with a grunting noise retreated back into the woods. Albert was not in any position to get a second shot. As we explored the site where he had fallen there was signs of blood. Al was disappointed of not making a direct hit and only wounding this animal. As we walked around we could see the direction which this wounded animal had taken. We proceeded for a considerable distance into the woods hoping to spot it but without any success we returned to our boat to continue with our trip; I could now see the river was flowing in a southwest direction. This was a clear indication that we were now on a return trip back to base. As we drifted along for another half hour or so we decided to would be a good time to stop for a tea break. We sat there drinking hot tea from steaming cups. Albert kept faulting himself for not using greater care in his aim in bringing down the second animal. I tried to reassure him it wasn't his fault and the moose was perhaps too far to cause a fatal blow. It was obvious by now that we would be going back home empty handed. We packed up, somewhat slowly now, and returned to the boat. The unforgiving river did not improve to any extent as we guided ourselves in a homeward direction. It was now getting later in the afternoon, so we were quite certain Bernie would be waiting for our return. There would be expectations that we would be carrying a load of moose meat. Other than keeping our eyes peeled for any further game we were very busy maintaining a safe as possible journey. Both Albert and I noticed the soreness in our arms from the excessive exercise we had endured along the way. We were completing a fairly sharp bend when we noticed some half mile down river an orange flag situated on the north side near the bank. This was the exit to the trail back to camp. The river at this point possessed a very strong current and we were moving at a rapid speed. We noticed Bernie and one of the other airmen standing on the shore waving towards us. Albert and I were using our paddles to slow down the boat to attempt a proper approach to the landing site. It was clear we were passing this spot so we attempted a sharp turn towards the bank. This move turned into a major catastrophe. The sudden turn resulted in our boat rolling over.

For the next few moments all hell broke loose. Albert and I were thrown into the frigid water along with all our gear. I seemed to tumble and roll about in this rapid current in total confusion. The river was about two hundred feet wide, too wide to get my bearings. I ended on a sand bar about on the middle of the river. There was a lot of shouting going on from the bank. I knew that if I stayed in the strong current I could work my way to shore swimming in an angle towards the bank. Once the boat turned I lost total contact with Albert so I feared for the worst knowing his medical condition. With a great deal of difficulty I managed to swim towards the bank. Bernie and his partner were on hand to drag me out. They told me Albert was okay, he had fought his way to the bank then continued to run along the shoreline with a view of retrieving the boat. The boat was just a short distance downriver stuck on a long jam. Although we wanted to help retrieve the boat, with both of us showing signs of hypothermia the airmen suggested the he take us back to base immediately for treatment. There were convulsive periods during the two mile drive back to our barracks. Upon our arrival wet clothing was discarded and we entered into warm showers. It was a reviving experience after our difficult and painful ordeal- to say it was pleasant would be an understatement. After donning a dry set of clothing we were able to settle down and review the circumstances of the ill fated day. The corporal's main concern was the loss of the military 303 rifle. That weapon was registered to this base with an assigned serial number. In the event it was not recovered he could received a severe reprimand. My 30-30 Winchester was also missing and a number of camping items which weren't as important. Supposedly the boat was still lodged within the log jam, so retrieving it would not prove too difficult. Bernie had already reported this incident to the Sergeant in charge and he approved for a recovery team to proceed to the site the next day.

Once I retired to my room I felt total comfort, but the cold shivers still plagued my body throughout the night. I was scheduled to go on day shift in the morning but made plans to change duties with Joe to give myself more time to recover. Shortly after coffee break the next morning a bunch of the fellows geared up and we were off to the river. Albert felt very guilty for this accident, and he was part of the team. We were both very thankful the accident did not happen somewhere miles up the river. If it would have, the result could have been disastrous. For the recovery mission there was a small boat with a kicker loaded into a truck, and a load full of other equipment they felt may be necessary in this case. There were a couple of wet suits available with Albert using one of them. Albert was an excellent swimmer in his day so he agreed to do some diving if need be. Once we arrived at the site, the first item on the agenda was to get a rough idea where our boat turned over. Bernie and his partner who were on shore when the dump occurred outlined the approximate spot. The small motor boat was set afloat with three aboard. As they manoeuvred into position they noticed their boat was quickly carried off due to the very strong current. A second attempt resulted in the same manner. It was then decided that a rope be strung across the river and secured at either end. A very long rope was on hand to conclude this plan at which time they were able to control the movement. With a rope also tied around him Albert also entered the water. He made efforts to dive but was quickly bounced up by the current. He estimated there was about nine feet of water in this area of the river. The mountain water was very clear but also extremely turbulent, kicking up a lot of dirt on the bottom making it hard to identify objects on the river bed. As the boat was kicked about, safely controlled by the secured rope, Albert continued to make dives. Finally he spotted some silver cutlery from our camping gear which showed up clearly. This was a strong indication our guns should be in the same general area. Continuing dives by Albert proved fruitless so it was decided that grappling bar with hooks be used to drag the river bottom. After a number of passes the hooks produced the desired results. My Winchester was pulled up with a hook holding on to the trigger guard. Watching from the shore I was very pleased to see my gun. For Bernie and Albert's sake though I was more hopeful for the recovery of the military 303 rifle. We did not wish any complications which could occur if recovery was not accomplished. As the dragging continued all we could do on the shore was watch from the sidelines. Then, all of sudden the drag line was being pulled in yet again and to everybody's surprise a gun was made visible. It was the military 303, hooked to the grapple by a shoulder strap. There were cheers of happiness by everyone as our rescue operation was accomplished. It was the feeling of the RCAF supervisor that it would be a waste of time and effort to attempt recovery of the remaining items lost in this watery grave. The small motor boat soon made its way to the log jam. There was no apparent damage to the boat either, so it was towed back to the landing site for loading.

The whole crew made our way back to the base for a late lunch hour but we were very pleased with the days work. I went back to work on the afternoon shift and took along my Winchester. It required a cleanup and some gun oil needed to be applied. Albert came to visit me at work and he felt much more relaxed after going through two days of traumatizing experiences. We talked in great detail about this river trip, where we could have done things differently. One thing we agreed on was we were very thankful about our survival. The moose we did not get was the furthest thing in our minds.

In the weeks that followed there was very little out of the ordinary happening. In my spare time I continued with my studies. After reviewing the various courses assigned in the flight services program, Jim felt I should have prepared myself sufficiently to write the exam. Almost a year had passed since my arrival in Smith River, and I was constantly expecting word about a transfer that was promised to me. Sure enough, one afternoon a teletype message was received. It was with great joy I finally got the news I had been hoping for. My transfer was approved, and I would be heading to Fort McMurray. It was to be effective the earliest possible date. Jim felt I could leave within the week. Headquarters was advised accordingly with a foot note that I was prepared to write my flight service exam. In response I was informed that in order to write my exam it would be forwarded to Fort Nelson. Enroute to Edmonton I would stop over there and it would be supervised by a senior aeradio staff member. Word of my transfer quickly spread around the base. News of a transfer, or any news, was a good excuse for a stag party. I do remember getting quite looped at my going away party which was something that did not happen too often. As had been the case leaving Embarras Portage, I was leaving Smith River with mixed feelings. Throughout the year I had been stationed there I'd developed friendships with a lot of the staff so there was some sadness in my leaving. On the other hand, living for almost three and half years in isolated conditions has done little to improve my social life. I was looking forward to the chance to return to busier locale. When the day arrived to leave Jim drove me halfway to Watson Lake where I was picked up by another aeradio staff member to catch my flight. Jim was the fellow I would miss the most, as he was always pushing me to improve myself. If it wasn't for him I probably would not have made an effort to write my exam. Upon my arrival at Fort Nelson I was greeted by the officer in charge, Stan Bilski. I was required to overnight and would be catching the CPA flight to Edmonton the following day. Arrangements were also made to write my exam the next morning. I booked myself into a local hotel. It was pretty rundown and the attached restaurant did not shine either when it came down to dishing out meals. I retired to my room where I reviewed my notes and hit the sack early. By eight o'clock I was in the radio office enjoying some coffee and chatting with the two fellows on shift. Soon after that, Stan arrived and after a briefing I was in his office writing the long awaited exam. This exam covered a large scope of subjects relating to the flight services area. The exam was finished before the 11 o'clock deadline. With this now behind me I was glad to be on the afternoon flight to Edmonton. Before proceeding to Fort McMurray I would be taking a couple of weeks of my annual leave. I spent the night with my sister Ollie in Edmonton then excitedly left to visit with my parents and family in Vilna. It was a very pleasant visit after not seeing them for almost a year. In two weeks my journey would be taking me to Fort McMurray.

Fort McMurray Posting

I was very happy to finally receive my transfer to this northern town. I'd passed though Fort Mac a number of times while stationed in Embarras Portage and was quite familiar with what it had to offer. After being stationed in two isolated locations back to back, this move would provide the social life which I had missed a great deal.

Upon arrival at the airport via C.P.A. I was greeted with a vigorous handshake by Vic Zariski the officer in charge. Being short staffed, Vic was more than pleased with his new arrival to fill the vacancy. The aeradio and weather office were located next door to the Canadian Pacific Airlines terminal. Both buildings were fairly run down, showing a lot of wear and tear, but were adequate. The runway still had a gravel base but there were signs work was in progress to complete the paving. Quickly I was taken to the Aeradio office and introduced to the fellows who were on duty. Generally two operators handled the day shift, one at the wireless position while the other handled the radio/phone. There was also a weather man.Vic introduced me to these fellows, Jim Kitagawa and Fred Burwell of aeradio and Bill Stern at the weather position. I already knew Fred, being we went through radio training together in Calgary. These fellows were single and lived in barracks which I would be assigned to. My new home to be was half mile away on the north side of the runway, next to the married quarters. After some chit chat, Vic drove me to the barracks for a room assignment. The barracks were fairly new, cozy and seemed comfortable. My bedroom was a reasonable size and close to a common washroom with showers, and a well equipped laundry area. These facilities were located at the west end of this long building. In the middle was a large common lounge room and the kitchen facilities which were located at the east end of the building. After dropping off my suitcase and duffle bag, I was taken to the kitchen area to be introduced to the cook. Bill Mitchell, a pleasant middle aged fellow neatly dressed in a white shirt and slacks pleasantly greeted me. He was known to prepare top of the line meals, but did not tolerate any form of inconsiderate behaviour in his mess hall. Bill had a room next to the kitchen but also had small house in Waterways, a suburb of Fort McMurray. He shared this home with his wife Dorothy. With Fred acting as a personal sidekick it did not take me long to get acquainted with everyone on base. It only took a couple of days to get orientated into the new position and I was soon able to take a shift by myself. My first impression of the fellows at the base was they were a nice bunch of guys to work with. Fred had a car so I joined him on trips to town where there was plenty of activities going on. My social life began to quickly fall into place. Being without a car was definitely a handicap since the airport was seven miles from town. I somehow got to know a few gals in town so life in general was treating me well.

Meeting Annabelle

A couple of the radio fellows who were married lived in the married quarters near our barracks. One was Ed Burrows, a fellow I also went through radio training with. The other, who would become an important part of my life, was Oscar Adami. He was a radio operator and agent with CPA and had been transferred to our Department only a couple of months earlier. I'd heard that he had teenage daughter by the name of Annabelle. It was well known that she was going out with a fellow named Hank. Hank was a summer U of A student who worked on a surveying crew with runway upgrading. I had casually seen her around with the impression being she was a nice looking gal. Since she was going out with this fellow I did not pay a great deal of attention nor was I introduced to her. It was after one of the dances in town that I asked to hook a ride back to the airport with one of the surveyors who were driving a large panel truck. The panel was full of airport people which included Annabelle and Hank. It was at this point that I was officially introduced. In the days that followed I had the opportunity to run into Annabelle occasionally. With each meeting we would have some brief conversation. She seemed to be a fun loving person and I kind of took a liking to her. Living at the airport Annabelle knew everyone there. I noticed that occasionally she would drop into the aeradio weather office knowing that I was on a particular shift. Working with her dad kind of made it easier to break the ice. Oscar would talk about his family and mentioned that Annabelle was only 16 years old and attending high school in town. He and his wife Mary also had a three year old daughter naned Helen. I was quite a few years older than Annabelle so our relationship was questionable. Off and on we continued to run into each other and for some reason there appeared to be a spark between us. Annabelle continued to go out with Hank, and I got to know and go out with a couple of gals in town. We all still attended dances in town, and Annabelle and I seemed to make a point to dance with each other. There was definitely a connection between us, but I did not push matters too far. It was sometime later she hinted to me that her dating Hank was nothing very serious. This got me thinking about things. For awhile after I did not seriously pursue anything relationship wise. However, during one of her visits to the radio office I took a chance and asked her for a date to a movie, which she happily agreed to. We had a grand time, but then let things ride a while. Then came a second date and obviously a closer connection was taking place. Little by little we kind of started going out on a steady basis. She dropped Hank and I stopped taking out the gals in town. Even though Annabelle and I came from totally different backgrounds, we seemed to have many things in common. As well she had a very outgoing manner about her, a mind of her own, and a very mature attitude for her age. After awhile I was invited to the Adami home for supper and had a feeling they had accepted me and approved of our relationship. Our biggest handicap was our inability to travel on our own. We had no means of transportation to get around but good fortune would soon change that. Bill Stern, a weatherman in the office had resigned his position and made plans to move to Australia. He had a small car he wanted to sell and asked me if I was interested. It was an English make Austin "A" which is a small car. It was only five years old, in excellent shape and the price was four hundred dollars. I didn't hesitate in coming up with the money. My bank account was slowly recovering after buying the farm in Vilna. Annabelle had received driving instructions from one of the weather fellows at the airport, so was excited about the fact we would have our own vehicle to go to town at our own convenience.

Christmas Visit, and Annabelles family

Christmas was approaching and I had requested a week of annual leave to go and visit with my family. I had not been home for Christmas for a number of years. Annabelle was disappointed I would not be around for the holiday festivities but understood. That year there was a huge amount of snowfall around Vilna. Traveling there from Edmonton required a great deal of skill on the roads. At least ten feet of snow blocked roads and single right a ways were common on the country roads. It was definitely one holiday season that stood out in everyone's mind because of all the snow in the region. After a lovely Christmas I was glad to be returning to McMurray. Excited about the fact I had some one waiting for me. There appeared little question in my mind that our love for each other was mutual. Annabelle continued to attend high school in town as I worked the various shifts. We would see each other usually on a daily basis. With a car at our disposal we were able to take in a lot more of the activities in town. Annabelle loved her little sister and frequently babysat. Often we would see a movie and Helen would be there with us, usually falling asleep. Annabelle referred to our cook as Uncle Bill, who was a good family friend of the Adamis. He was kind of protective of her in the beginning, but now knowing me fairly well he saw I was a pretty nice fellow, so he approved of our relationship.

During the time we spent together Annabelle often spoke of her family and the many events that had taken place in her life. She was born in Whitehorse, Yukon and had lived with her parents on a trap line located on the Pelly River until the age of nine. Her family were well known "Yukoners". The great-grandfather on her dad's side went to the Yukon during the gold rush at Dawson City. Her grandfather who had lived in Vancouver had also moved there after his wife passed away at a young age. Her dad, Oscar, at that time was only twenty one years old and an only child. Ann's mom and dad met while working on a tobacco farm at Samas Prairie in southern B.C. They were married shortly after, and Oscar journeyed to the Yukon as well to seek a better life. Mary had stayed behind until he had earned enough money to have her join him. She did so a few months later but their lives were far from easy. They both cut firewood for the steam boats along the Yukon River to earn enough to live. In Fort Selkirk through a local RCMP officer the young couple learned of an available trap line some seventy five miles east on the Pelly River. The trapper who owned the line had disappeared leaving a suitable cabin, traps and other gear used for trapping. The total price for everything was only $100. They agreed to purchase this line, but stayed the reminder of the summer in Fort Selkirk cutting firewood to pay for a grub stake. That was in 1936. Oscar and Mary were avid and successful trappers and earned a good living off the sale of furs they caught. The winters were spent on the trap line and returning to Fort Selkirk to a summer home they purchased.

When Annabelle was nine the family had to leave the trap line because their line was located in an area known as the McMillan basin and was declared a park. The family moved back to Vancouver where Oscar enrolled into a commercial wireless radio program at the local college. He earned his second class license and was hired by the Canadian Pacific Airlines as a radio operator in Fort McMurray.

While living on the trap line Annabelle was homeschooled by her mother, so there was a bit of culture shock when it came to attending public school in McMurray. She did eventually adjust to life in the town and continued to attend high school during that early time in our relationship. I found working the station more pleasant as well. I'm sure it had something to do with Annabelle, and having a steady girlfriend to spend time with. I also found the officer in charge and the rest of the staff to be a great bunch of fellows to work with. As winter had settled in the usual outdoor activities slowed down. I found the workload on a shift considerably busier then what I was used to in the past. The volume of air traffic in and out of the airport was largely related to a large scale mining operation being carried out at Uranium City and another at the Stony Mountain radar site, which was being built twenty five miles south near the hamlet of Anzac. I loved living in comfortable barracks in McMurray. Uncle Bill dished out very excellent meals, and as usual I had to use good judgment to keep my weight down.

In April Oscar booked his annual leave and Mary and young Helen left on their holidays to visit some very close friends who lived in Powell River B.C. This was a small coastal city with a large pulp mill. During their three week visit to that site they fell in love with the area. Oscar made an inquiry at the Pulp Mill and they were in need of electricians. As Oscar had a good background in radio of which electrical knowledge was a big part; he applied and was successful in getting a job. He had agreed to complete training for a journeyman's ticket as well. Little did I know the big impact this decision would have on my future.

Adami Family Move to B.C.

Returning after his leave to Fort McMurray Oscar put in his resignation. He gave a months notice. This threw a monkey wrench into my relationship with Annabelle. She would be moving to B.C. with her parents. By now our love for each other had blossomed and it would be a sad time in our lives once she left. We had already discussed marriage and without much hesitation I purchased an engagement ring. She said yes. Ann's parents surprised us with a small engagement party. I told my parents what had transpired and they were happy to have a wedding at the farm. As I was the first in the family to get married they would certainly go all out. Ann and I were excited and nervous to move on the next stage in our lives together, with so much about our future still unknown.

Despite the happy news of our engagement it was a very sad day when Annabelle left with her parents. She would return about mid July with our wedding date set for July 28, 1956. The next three months apart was a difficult time for both of us. We wrote letters to each other daily which eased the loneliness. Annabelle had a job at the beach and was getting a good tan. I kept busy at work spending my free time with the fellows in barracks which I had neglected in the past. I heard from mom and dad they had started sprucing up the buildings and yard for the wedding. As the loft of the large barn was to be used for dancing no hay would be put into that area. The lower part of the barn was also to be cleaned and whitewashed. The garage would be used for serving meals so it also had to be cleaned properly. Annabelle's list of people attending the wedding was not very large nor was mine, although it was a family tradition for relatives, neighbours and friends to be invited. This meant about three hundred people.

Embarras Portage: Before, and After

I was never so happy to see the middle of July arrive. Annabelle would be arriving by bus on July 17. I would be on annual leave and meeting her in Edmonton on that date. Working with the CPA pilots on a daily basis I got to know most of them. They were also good friends of the Adami family having worked with Oscar in the past. Knowing my happy destination the captain of the DC3 flight I was catching told me not to bother to pay for the fare. He told me to board the aircraft with the rest of the passengers and I would occupy the jump seat in the cabin. I guess this was his wedding present to us.

Our Wedding & Honeymoon

Seeing Annabelle after all these months was a real joy. We planned to hang around Edmonton for a couple days to tie up some loose ends and finish planning for our big day. We had the good fortune in staying with my sister Ollie, who was also being Ann's bridesmaid and was helping her out with some of the necessary bridal things. We proceeded to the family farm on the weekend before our wedding and were greeted with a huge homecoming party. Annabelle was kind of in shock with all this activity. First of all meeting my parents, siblings and dozens of friends for the first time. She had never known or been exposed to a Ukrainian traditional type of gathering such as that. Looking back I have to say she did manage quite well; it helped knowing I was close by to support her.

Our wedding, long in the planning, was a delightful affair. Mom's neighbour ladies spent a week helping prepare the food for approximately three hundred people. Dad made sure that there was plenty of beer and whiskey on hand. Brother Mike and brother-in-law to be Ed spent some time in the muskeg making a brew of moonshine, which went over really well. Family and friends spent two days enjoying this special event. The only very sad part of our wedding was Annabelle's dad was not able to attend. As he had only recently secured employment at the pulp mill he was unable to get any time off. Ann's mother was able to attend, one of the few family members who could which was a great comfort to her.

A few days after the wedding we were off on our honeymoon. We booked a few days off for a small vacation via CNR train to Vancouver. There we spent some time visiting with Ann's relative including her grandparents on her mom's side. After a couple of days in Vancouver we boarded a bus to Bellingham, Washington. We stayed in a nice hotel, taking in movies and sight-seeing. Then it was back on a bus through Vancouver on the way to Powell River. As this was my first trip to this part of the country I found it very exciting. Ann was overjoyed to be with her family again. I had taken only three weeks of my holidays and we spent only a week in Powell River before heading back to Fort McMurray.

Back in Vilna we had a brief visit packing all our wedding gifts. We were then taken to Lac La Biche where we boarded the NAR passenger train, known as the muskeg express. The stopover in Lac La Biche from Edmonton gave everybody the opportunity to purchase a good supply of booze for the long trip north. We would be boarding about 2 pm in the afternoon and not arriving in McMurray until about 5 am the following morning. We purchased a sleeper, but never had the opportunity to use it. It seemed that a great number of the passengers were party people. There were plenty of drinks and people with instruments ready to drum up some singing and dancing. Due to the excess noise the party bunch was sent to the baggage car. Once it was made know that we were on our honeymoon the drinks were free. With no sleep and the entertainment continuing all night we were not in very good shape by the time we arrived at our destination.

Much to our surprise, a homecoming party was in the works for the evening. So you can only imagine how alert we both were to attend this special event for us, but we gave it our best shot. The party was held in the lounge area of the singles men's barracks. Almost everyone from the airport plus a few close friends from town were there. It was a great time which went on to the early morning hours.

Back in Fort McMurray

After our marriage I was no longer a resident of the single barracks. Prior to our departure I had made plans to rent a small dwelling in the town waterways. It belonged to the cook and his wife. It was a small, but comfortable two bedroom home. The only thing that it did not possess was indoor plumbing. Water was delivered in by forty five gallon barrel which was common for the town. This house was furnished with most of the basic needs so we didn't have to buy any furniture. It had an oil stove and heater as well. One inconvenience was having to make the drive to work now, a distance of seven miles along a narrow sandy road that had a total of ninety-two curves.

As we settled into a married life, we enjoyed ourselves during this exciting period with some occasional minor ups and downs. At first Annabelle had little knowledge when it came to cooking, so we worked on this together. There were many burnt pots filling up in a wood box outside which had to be cleaned up when the supply ran out. Frequently for a good meal we would go to Mah's Café for Chinese food, and at other times we would visit Uncle Bill at the airport who would always offer leftovers. As time went on Annabelle's ability in the kitchen took a crucial turn that was most outstanding and remains to this day.

During the latter part of the summer we received notification that Vic Zariski, OIC was being transferred to Edmonton taking on a position in the technical field. Our new in coming chief was none else than Curt Hart. I was pleased as Curt was my trainer in Edmonton when I first joined the department. Curt, his wife Ruth and their four children soon arrived there after to take on his position. A close relationship quickly developed between the Hart family and ourselves. We visited frequently with each other and often when I worked the midnight shift Annabelle would spend the night at the Hart home as she did not like the idea of spending nights alone.

My brother Stan had written me stating the he was planning to purchase a new car. He knew that I liked this 1951 Plymouth and asked would I be interested in buying it. The little Austin was fine but we wished for something a little better, as well, I had a buyer for it at a good price. I quickly let Stan know we were interested in buying his baby blue Plymouth and we knew that this car was in excellent shape and his price was very reasonable. He had it loaded on a NAR freight car in Lac La Biche and it was shipped to us shortly after. That car was our pride and joy for many years after. Cars were few and far between in Fort McMurray and ours stood out well in the crowd.

Annabelle had quite a number of very good friends from her high school days and in addition to those at the airport we both knew, we carried on a fairly active social life. We attended dances and movies frequently, joined badminton club and on the lighter side would often play cards, gathered around friend's tables to stave off the cold McMurray nights. With the early part of winter settling in we were looking forward to our first Christmas together. To our surprise, we received a letter from my parents they planned to come visit us during the Christmas week. In preparation we located a nice small spruce Christmas tree. We purchased some ornaments and set of lights, and being it was our first attempt it turned out fairly well.

Parents Visit

My mom and dad, along with my five year old twin sisters Lucille and Laurette, arrived by train a few days before Christmas. The weather had turned very cold and was accompanied by heavy snow falls. As usual, mom was very generous in providing us with farm produce. She brought with her various types of meat, eggs, garden and a turkey for the Christmas feast. On the trip back to our home, Dad talked about how he had been talked into a low stakes poker game on the ride. Mom said she had found the trip uneasy because of the excessive noise and carrying on which made it difficult to get the little sisters to settle down during the late hours. It was a lovely holiday. After attending midnight mass we enjoyed a festive lunch. The next morning came with the exchange of gifts and another well prepared meal including the turkey and trimmings. A few of our friends dropped in for a visit as well. Mom and Dad enjoyed the few days break from doing farm chores as Frank was taking care of things at home. I sensed some sadness though in Annabelle, who was missing her family. I felt badly, understanding exactly what she was going through. My twin sisters played with the new dolls they got from us. Two days after Christmas my family departed for home. I was lucky to have a couple of days off during their visit. I went back on the midnight shift which also including working on New Years Eve.

In the north, New Years celebration is a very special event and plans were made for this year as in years past for the big party to be held in the lounge room of the single men's barracks. It involved all members of the airport staff and a number of the town people. It began with a big turkey supper prepared by Bill and assisted by the ladies at the airport. This was followed by drinking, singing, dancing and general fun. I was working the midnight shift and had to play it very cool in a way of celebrating. I went to work well before twelve in order to give the evening worker some time to join in and catch up with the rest of the bunch. Annabelle came in at midnight to wish me a Happy New year, our first together. She returned to the ball to enjoy the rest of the night. She was spending the night with the Harts.

The cold blast of winter continued with even more snow and temperatures well into the 40 degree below range. It was always a chore to keep the vehicles operating even though they were plugged in on a continuous basis. A few days after New Years I received some very sad news. My close friend Johnny Gushta had been killed in a train accident. The news hit home. Johnny and I grew up as neighbours living on a farm less than a mile east of us. He was my age and we shared a lot of experiences together. We'd gone to high school together but after high school we each went our own way. I went to radio school while he found a job as a locomotive fireman with CNR railway. Johnny got married very young and had a wife with two small children. From what I was able to gather he had just completed a trip getting off shift in the early morning. He lived not too far from the railway station and was heading home walking along the tracks. A pretty severe storm was blowing at the time. Bundled up against the biting cold he did not hear the train approaching which hit and killed him instantly. Being close, like brothers, this news upset me greatly, especially knowing his young wife was now a widow and his two small boys were without a father. Due to the very cold and unpredictable weather Annabelle and I stayed close to home keeping our small house comfortable and warm as possible.

Kitchen Accident

My birthday was on January 12th, so we talked about going out for Chinese food at the Waterways Café. Annabelle had stayed up the night before to bake and decorate a birthday cake for me. The day of my birthday Ann decided to make a seven minute icing for a special treat. She had looked up a new recipe in a cookbook. It was almost midnight when I heard a huge explosion from the kitchen followed by a loud, painful scream. I raced in from the bedroom and was horrified to see Annabelle trying to hold a large cloth over her face. Once this was removed, large pieces of skin came off from her around her face extending all the way down her neck. The burns were severe looking, and the pain and trauma were excruciating. I quickly had Annabelle bundled up, with her face fully covered from the harsh winter night. Luckily I was able to get the car started in the sub zero temperature, and it did not take very long sliding along the roads to get to the hospital. The nursing sisters quickly got Annabelle settled down, applying the necessary medication to her badly burn face. Dr. Pickford was called in to view the damage, which was still hard to tell at this stage. I also learned the reason for this catastrophe was a mistake anyone could have made. Annabelle was supposed to have used a double boiler pot for this icing recipe, but used two pots instead, one sitting on top of the other. Cooking on high heat there was a dangerous build up of steam in the lower pot, and with the two pots fitting tightly together the excess pressure caused the blow up, sending molten hot icing from the top pot flying up into her face. Annabelle spent almost a week in the hospital, and it took a considerable time to properly heal, but thank God there was not any noticeable scarring.

It seems that if you encounter one problem, it's often followed by another. A couple of weeks after Ann returned home from the

Embarras Portage: Before, and After

hospital, with the weather still running extremely cold, we experienced another near disaster. One cold night one of us woke up from a deep sleep to notice the house full of smoke. Quickly getting out of bed, we noticed the oil stove was giving off obnoxious fumes, causing sore eyes and breathing problems. I quickly knocked off the oil flow in the stove but being unable to remain in the house we were forced to quickly bundle up and walk up to our neighbours who lived down the hill. The Listers welcomed us into their home for the night after we explained our problem. The next morning we entered the cold house to find the cause for the smoke and fumes. Taking a look at the pipes and chimney it was clear there was a great amount of debris blocking off the brick chimney. Upon clearing out the debris and making some repairs, I relit the stove and a few hours later our house was back to normal other than for a lingering, unpleasant smell.

After a difficult and cold winter, it was with great relief that we welcomed the spring season. The Fort McMurray area was fairly low key this time of year, with population remaining in the neighbourhood of 700. As well the boat crews would be moving in to work at Northern Transportation and other projects, bringing an additional 500 people who would make this their summer home. We also noticed an increase in air traffic with most flights going into Uranium City, where mining for uranium was a major operation.

During the late spring our office received notification that Army Signal corps was leaving the northern postings. Their extra duties and facilities were being taken over by our department. Fort Smith NWT was then the capital of the Northwest Territories, with the Signal corps housing a large staff. Now being a fairly senior member, I was asked if I would consider a transfer to that northern town. One of the perks would be a brand new three bedroom house, fully furnished with only a small basic rent including all utilities. Annabelle and I discussed this and we agreed it to be a good plan, and I submitted our application. At this point in our lives we had few possessions which included only our personal items and the wedding gifts we had received, so a move would be an easy affair.

Fort Smith N.W.T Transfer

Our transfer was approved and in early May we boarded a DC3 regular flight to Fort Smith. A few days prior to our departure the airport staff held a big farewell bash in our honour. By the attendance it was very obvious that we were leaving many good friends behind, but we wouldn't be too far away.

Upon our arrival in Smith were met at the airport by my new officer in charge, Kevin Coyne. I knew Kevin well as he too was one of the first persons I had contact with when I joined the department. The middle aged, grey haired gentlemen greeted us warmly. After our small amount of baggage was loaded into the pickup he took us to our new home. It was on the airport residential site half a mile from the terminal. Our new home was in a newly built duplex in a compound close to a variety of government buildings, some older and some new. The whole yard lacked any grass, and consisted of fine sand with no signs of any landscaping. Kevin presented a set of keys and we entered our new home with a great deal of anticipation. It was full of new furniture- so new, that most of it was still in crates. It was very obvious that there would be a great deal work required to sort out all the mess. Before we dove into the task of settling in, Kevin invited us to meet his wife Lil. She had prepared a lunch to welcome us to our new neighbourhood. Ann and I could see she was a warm, soft spoken and caring lady. We learned that they had ten children ranging in age from sixteen to down to an infant. Kevin drummed up some help with unpacking and placing furniture. We could not believe that large size and condition of our new home which was something neither of us had been used to. One of the few things I remember we did have to struggle with was mounting the new blinds and hanging curtains. All in all our first few weeks in Fort Smith were pleasant ones. Little by little we got around to meet our new neighbours. Many of them were young families like ourselves. Their duties varied from aeradio staff like me, to weather observers, power house operators and maintenance staff. Fort Smith was a major airport so a considerable amount of air traffic was passing through on a daily basis. In addition to the married quarters, there was a single mans barracks also located in our compound. Only recently our department completed a total takeover from the National Defense Department, the Army signal station and it's accompanying local resources. This included five single dwellings for married staff, and a large married barracks with total kitchen facilities. There were also a number of storage facilities had housed food supplies and rations for the airport staff. The army signal corps main duties of the station had been to disseminate government and commercial correspondence by means of wireless radio. Fort Smith was a main staging area and collecting point for all southern traffic being forwarded to a central location in Calder (Edmonton). Being headquarters for most federal government a large amount of traffic was handled there on a daily basis. The RCMP housed the "G" division had a small prison which had once even hung criminals who committed serious offences such as murder. All Mounties who had signed up for northern duties took all their training here. As dog teams were still used extensively than, they were required to train heavily to become proficient sled runners. Due to the transfer of the station from one department to another, a considerable amount of extra staffing was needed, as well as some in-service training. Aeradio staff at the airport was increased from the regular nine members to a total of seventeen; which now included the Army Signal office. All operators were trained to work at either location. I came on staff as second-in-charge next to Kevin, so I had additional administrative duties placed on me, mainly coordinating and preparing shift plans as well as monitoring an ongoing audit of food supplies at the single men's barracks.

We had been married for over a year now and it seemed that everybody was starting to wonder when it was going to happen. Having children had always been our expectation, and Annabelle and I had talked alot about having a family. Finally, the good news came. We were expecting, with an arrival date in early November.

Annabelle did get checked out by a young doctor recently arrived in Smith and everything was normal, plus the fact she was feeling fine.

Meanwhile I was busy acclimating at work, and getting to know the staff I was working with day in and day out. Most of the aeradio staff were single fellows on their first posting. There were two chaps from Newfoundland, one who was a very unusual character. He was brought up in a very remote fishing village on a coast line where his father was a fisherman. He received a government sponsorship to go in commercial wireless training. He was assigned to our station once he completed the training courses. David Winsor was extremely backward, and had little knowledge about looking after his personal needs, even things as simple as purchasing items of clothing or how to make an appointment with a dentist. He spoke in a very unusual fashion and his speech was at times hard to understand. This created difficulty at the station with his communication with aircraft, having a hard time clearly giving the pilots certain information. On top of that, it was also David's first time away from his family and he was extremely homesick. His mother had always done everything for him, so he found it difficult to adjust to a more independent lifestyle. I kind of took this young lad under my wing and helped him in any way I could. As time went on there was a great deal of adjustment on his part, and his situation slowly improved. Another very unusual fellow was by the name of Tom Anarluk. He was of Eskimo decent from a place called Coppermine, NWT. He was orphaned at a very young age and brought up at a Catholic mission. Tom was a very bright young fellow, and after completing high school received a government sponsorship to enroll in a commercial wireless technical school in Toronto. By selecting this type of training he knew he would be able to return to the north to secure employment. He was stationed at a few northern locations before being placed in Smith. Tom had a very outgoing personality, and was very popular with the young single ladies. He loved to party and had no concept of money, living from one payday to the next. On one occasion he acquired a total of seventeen pairs of jeans and almost the same number of shirts before deciding to have them taken in to a laundry. The story went that after buying a pair of jeans he would wear them until they became too dirty, lay them aside and buy another pair, leading him to aquire a very large wardrobe.

As far as the rest of the people on base, many had young families such as ourselves. Living in a close community setting we acquired many good friends. We soon learned that living in an isolated northern environment brings people together. We all were very closely knit and shared in common goals. On the lighter side we would hold card games followed by steak fries, usually buffalo during the cold winter months. We would have house parties pretty well almost every weekend which included dancing to a record player with Johnny Cash as the most popular LP. Annabelle and I loved to dance. With a lot of practice we become very professional in jiving. After we would set the old dance floor just a waving in steps with the rock and roll records. Even with the now increased size around the tummy, Annabelle could still cut a mean rug, but was careful about not over exert herself and knew her limit. Most of us drank beer, which was usually homemade. Each person would take turns preparing a batch and when ready, the party would be in their home. Generally these house parties were in good taste and not wild or disorderly to any extent. We would hold a "do" in the recreation area of the single men's barracks. Our department established a very good rapport with the RCMP , since we handled most of their correspondence to outposts in the north. Most of the young Mounties were waiting placements to various small detachments but first underwent a period of orientation in Fort Smith. They had a very restrictive lifestyle in the barracks, living with a sergeant who was a real disciplinarian. On their free time, which was very limited, they often came to visit families at the airport and also often joined in some of the many social events that went on at our base. One of the young constables we got to know was Dave Hawker. He was assigned for northern duties a couple of months previous. Dave was from Toronto. He told us he was engaged to a young lady but that a future marriage was still a long way off. The RCMP policy of that period was a member had to be in the service for seven years before he could get married. Dave visited our home whenever he had some free time. As

Annabelle's cooking was wonderful, he often had dinner in our home as well. Being a very good looking lad he didn't have any problems in getting some the local gals to give him a chase. He soon developed a steady relationship with a blond gal who worked in the post office. Also during this time both Annabelle and I developed a very close relationship with Kevin and Lil Coyne. Lil had given Ann a great deal of advice and comfort dealing with her pregnancy, being that Lil had ten children. The Coyne's had close ties to the Catholic Church and we often attended mass together. Every Sunday it was Lil's day off from any chores. It was Kevin's turn to cook a large dinner for the family. Very often we too were invited to attend a well prepared meal. Kevin would bake buns and a number of pies. Usually it was a big roast with vegetables and all the trimmings, followed by a dessert. Mr. Coyne was Annabelle's trainer when it came to bake bread and buns. Many people often wondered how this family managed in feeding such a large family in the north. The Coyne family went on holidays in July for a whole month to their cabin at Ma-me-o beach. Just before their return they would attend a whole sale in Edmonton and other stores, spending thousands of dollars in groceries and meat. The order would be shipped on one of the last NTCL barges from Fort McMurray. This is the kind of planning required to feed a family of twelve in a place where fresh foods could often be in short supply. Another family friend was our next door neighbours in the duplex, the Macleods. Norm and his wife were a number of years older than us and they did not have any children. Norm was ex-military type serving with the Navy as a wireless radio officer. After five years he did not wish to re enlist and ended up with our department. Norm had a very outgoing personality, while his wife was a very quiet and mild mannered lady. He was one of the main organizers when it came to brewing beer, and he taught us a lot of the skills to brew our own. The normal batch usually consisted of three dozen bottles which was about the right amount for a party. Norm loved to party, and usually ended up in playing his bagpipes of which he was a master at. He was very proud of his Scottish background.

    Norm was also an avid fishing man and he had done a little research on fishing for lake trout at a nearby lake just north of us.

One time, as it was getting close to fall he decided to organize a fishing trip to this isolated lake forty miles away. We would charter a Norseman aircraft flown out of Smith by Associated Airways. There were six of us in the group that agreed to be flown out for the day. There was Norm, myself and Steve, another fellow from our department. Three RCMP fellows joined us in this expedition. It cost us seventy five bucks a piece and we were off one very early morning. We took along some cooking gear, hopefully for a good fish fry. The time from take off to having this lake in site was less than an hour. The terrain was sparse in areas, solid growth of spruce or jack pine forest in others with a lot of muskeg and creeks in between. The pilot made a pass over the lake which appeared to be two to three miles across. As the pilot had dropped off fishing parties here in the past he knew a suitable area to land and taxi for easy exit. The Norseman manoeuvred close to the shore and one of the fellows jumped out on the float and with a long rope to secure the plane to a nearby overhanging tree. After unloading all the gear the pilot started up the motor and shouted over his shoulder that he would be back for pick up about 4 that afternoon. It was only a little past eight when he dropped us off so we had the better part of the day to do a lot of fishing. It was a beautiful lake, remote and wild. The water was deep but appeared crystal clear. The area where we left the aircraft appeared to be a regular of a drop off point for other adventurous travelers. There was a sign of a camp out nearby. Along the shore line ran a long sandy beach. I could only imagine living near a lake such as this if it was easy to get too. Before getting down to any form of a serious fishing the first item on the agenda was to build a fire and brew some coffee, one of the best parts of any wilderness trip in my mind. After belting down a couple of cups together with a few cigarettes we all got serious about what we came here for. We gathered our personal fishing gear and walked along the shore line to locate a suitable site for casting. In a matter of a few minutes some of us were lucky and began hauling in some beautiful specimens of lake trout. We casted, knowing if we did not get a strike there was a good chance of losing a hook on a snag or gravel bed. The smaller fish we brought in were mostly released knowing there would be no problem

Nicholas Verbisky

in getting our quota before our flight returned. We took frequent coffee breaks as the fish were coming in a lot faster than expected. With the approach of the dinner hour we all stopped to fillet a number of these price catches for the noon meal. We all brought along frying pans and other necessary items for this occasion. To my recollection I can't recall a fish feed equal in anyway near to the one that day. There was no set limit on the amount of fish we could catch or eat and other than a few muskrats and beavers swimming around occasionally, there was no other sign of life in the area. At about 3:30 in the afternoon we heard the approach of our aircraft and we were ready to head back home, each one of us packing a sack full of this fine fish to take home with us.

Fort Smith N.W.T

We had a fairly large yard near our duplex, but it was solid sand where little of anything grew other than a few lonely weeds. Most of the residents paid little interest in the appearance of their yards. Annabelle liked growing flowers, while I had always been interested in a vegetable garden, growing up with one at the family farm. I took a serious interest in making the yard more suitable for growing but there was little, if any, suitable top dirt in the area. I located some suitable material near our transmitter building. It was scraped up by a caterpillar when clearing of the site was concluded. Using the department pick up I hauled in a number of loads. It was too late in the summer to do any planting, so I wanted to prepare the area for the next spring.

After living in the area for a number of months one finally got their bearings. This large government town was capital of the N.W.T and was unique in many respects. Isolated from the outside world, with the only access by air or by river, it was pretty well self contained, located about seven miles inside the N.W.T border. The population at the time was in the neighbourhood of 2500, but varied depending on the time of the year. Fifteen miles to the southeast was the small hamlet of Fort Fitzgerald. A good road connected these two locations. Fort Fitzgerald was located on the Slave River and a drop off point for northern transportation cargo going to Fort Smith and other northern stops, due to the rapids located near Smith River. Any other traffic through that area was not possible. All goods and supplies had to be portaged and transported by road between Fort Fitzgerald to a point beyond the rapids, a distance of about twenty miles. From there, cargo was reloaded on barges for all points north on the Mackenzie River. Fort Smith was located near the Slave River with the roar of the rapids ever present. The layout in itself was very scattered. The greatest number of buildings belonged to the government departments. There was our compound, former Army Signal corps. Four married quarters, a signal station located on the bank near the river, a number of storage shelters and the single men's barracks with kitchen facilities. The RCMP had a large setup which housed "G" Division headquarters, training center for northern recruits and accommodation facilities for the air division. Other federal department such as Northern Affairs, Housing, Parks and Fisheries also maintained facilities and staffing to serve all the Northern communities. There was the Hudson Bay Store and two other, smaller ones which carried mostly food items. Due to the fast expiry dates, perishable items had to be flown in, and were terribly expensive. About half of the residents were natives, many who made a living by trapping in the winter and were involved with fishing or working with the forestry department during the summer months. The St. Ann's hospital served all the medical needs in the area. It was operated by the Catholic Church, with the majority of the nursing staff being sisters. The Catholic Church had its main base here which included the Bishops headquarters for the NWT. Getting around could be a challenge. There were only a few main roads in the region including the portage road between Fort Smith and Fort Fitzgerald, the drop-off point west of town. The four miles of road between the airport and town was well traveled and maintained. There was also a network of hiking trails throughout the area. The land mass was very sandy in most areas, making trail building relatively easy. This was handy, as the entire region was in the Wood Buffalo National Park. Annabelle and I became very friendly with a couple who lived in a duplex a few doors down- Jerry and Mary Bourget. They were close to the same age as us and had one small child. Jerry came from Lac La Biche and was working as a maintenance man for the Civil Aviation department. Coming from work I was never sure if supper was in either their house or ours. We were also very happy having our 51 Plymouth which we had shipped from Fort McMurray shortly after our move. Often on days off our two families would load a picnic lunch and head down one of the trails. There were the salt flats, which covered hundreds of acres of prairie looking landscape.

Growth of the few trees was stunted due to large amounts of salt in the soil. The area was frequented by large herds of buffalo. We would drive right amongst these large animals, only later being told how dangerous a chance we had taken. We easily could have been run down by one of the one ton bulls. Another site we often visited was Pine Lake, 25 miles southwest which we reached on a rough and narrow dusty trail. It had a nice campsite and the waters boasted many jackfish which we usually enjoyed, depending on our luck.

Being second in charge of a pretty large establishment I enjoyed some privileges, with the trade off being some extra duties to perform. Most of the time I worked regular shifts like the other aeradio operators, but sometimes I worked the straight day shifts. I would also assist Kevin in making sure operations were running smoothly. Maintaining an accurate shift plan for seventeen staff created problems at times. Keeping proper maintenance for our equipment was an ongoing job. Not to mention promoting a harmonious relationship amongst the operators. Many of them were young and inexperienced, and the isolated lifestyle did not make it any easier. Personal problems had to be handled with a lot of compassion.

Anthony Arrives

We were anxiously looking forward to the arrival of our first born. Annabelle visited with Lil often, and got frequent updates on delivery matters from the very experienced mother of 10. On the evening of November 2 we were visiting with the Coynes, and Annabelle was showing signs that indicated the baby may soon be arriving. Lil instructed me to get the gear together and head to the hospital. After getting Annabelle settled and comfortable the nursing sister in charge of maternity suggested I go home to bed as it didn't seem that baby would arrive that night. In any event men in those days did not have access to the maternity ward, or were allowed to be in the delivery room when the babies were born. Early the next morning I headed to the hospital to determine if any progress was being made. Much to my surprise, I was informed that I was the proud father of a healthy baby boy. Apparently the maternity nurse had misjudged and the delivery happened only a couple of hours or so after I was sent home for the night. I was glad to hear that Annabelle had a fairly easy delivery and we were both excited about the new addition to our family. We had already agreed on a name, Anthony Nicholas, which was the first name of my paternal grandfather. This new bundle of joy would be an adjustment to us both, being inexperienced parents. Annabelle spent the next five days in hospital, which was normal for new mothers of that time. This little fellow seemed so fragile when I was holding him at first, but slowly the fear of causing him any harm diminished. The day after Anthony was born we sent out radio telegrams to both Annabelle's parents and mine. Being the first grandchild for both sets of grandparents there was a lot of excitement expressed in the lovingly written cards we received. Getting him home was another experience. He was nursing and the feedings were very frequent which brought upon periods of colic attacks. These were handled with advice from Lil to have a bottle of Gripe water handy. After the first month or so we were much more at ease in handling our young fellow.

I had two weeks of annual leave still coming to me so we decided to visit both of our families for Christmas holidays. Leaving Fort Smith in full winter gear at a temperature of -40 degrees, changing planes in Edmonton and arriving in Vancouver where it was much warmer, and raining, was a considerable but welcome change. After spending the night in a Vancouver Hotel we were bound for Powell River the next morning. The week we spent with Oscar, Mary and Helen was wonderful, with little Tony being the centre of attention. I was scheduled to be back in Smith prior to New Years Eve, but we also planned to spend almost the same amount of time with my family before our return. In all fairness for having the opportunity to spend the Christmas holiday with our families, I would be returning to take on regular shifts for New Years Eve and New Years Day. This was to give staff that had been working during Christmas holidays time off to celebrate. We returned back to Smith on a scheduled CPA DC3 flight two days prior to New Years Eve.

Other than the regular flights in the area air traffic was very slow during the festive season. Every New Years Eve there was always a big bash held in the recreation area of the single men's quarters, one of the social highlights of the year. All airport residents were welcome to attend. Single staffs were allowed to invite dates. About half a dozen of the single RCMP fellows who were frequent visitors at the airport were also invited to come and bring along their girlfriends. Canned music was provided by a well stocked record player. The cook always made a special effort to prepare a variety of tasty dishes for the occasion. Everyone brought along their own supply of spirits for the party. A common drink was known as "moose milk", rum and eggnog. Our duplex was located only about a block from the single men's lounge. Annabelle and I planned on attending this very special evening. Little Anthony would be babysat the Coyn's fifteen year old son John. The party was hopping- well attended and the night progressed in a in a very festive manner with the usual dancing, eating and socializing. No one expected the sudden change in weather. The temperature was still hanging around the -40 mark. Than at eleven o'clock it started to snow heavily, followed by a strong north wind. In a matter of hours the storm became so severe we could hardly see across the street. With the previous snow on the ground and the new snow falling heavy, drifting began to occur. This put a damper on the party and some of the guests from town decided to head home as the snow began blocking the roads. As the storm continued even Ann and I decided to go home, and even the short distance we had to travel we also found ourselves bucking snow drifts. Our RCMP friend Dave Hawker, his girlfriend and a few of the other Mounties and their dates were invited to our home. They decided not to take the chance of being stranded on the road and would wait out the storm. There was still plenty of moose milk to consume, so a poker game was organized. I was going on shift in the morning and decided to call it a night, but the unplanned gathering went on long into the night.

There were a couple of the civil aviation fellows on standby to handle situations such as this. The snow plows and graders would be put into action to clear road ways first thing in the morning of the new year and once the storm subsided they would be out on the runways blowing off the snow. With day break of a new year, the nasty storm came to a near standstill. The road to the terminal was cleared so I was able to get to work, catching a ride with the night shift worker. Other than sending out a NOTAM notice that there was equipment working on the runway, things were very quiet. The weatherman on duty had some very unusual reports to submit regarding the storm.

Dave Hawker, one of the young Mounties with whom we became good friends, would visit with us during any spare time he had. He would often come to the airport to chat while on a training run with his dog team. The six huskies were a prize team, and I occasionally went on a trip with him. After my time in Embarrass, I was familiar in handling a team so sometimes I would take command and put them through their paces. From Kev Coyne, Annabelle learned how to bake bread and buns which was always David's favourite; a real treat he never received in barracks. Dave would often talk about the personal problems he was having. The story went something like this. He was engaged to a girl in Toronto but being away for such a long period of time he was lonesome for female company. He started taking out a nice girl he met in town. She was a friendly young blond and had come to our home with Dave a few times. She worked in the post office and was living with her family who were longtime residents of Fort Smith. Dave felt very guilty about what had happened and was not sure what to do. It was hard to see our friend going through what could only be called a complicated situation.

Following the New Years storm the weather was brutal for the next couple of months. Mostly it was very cold, with frequent storms with a lot of snow. The limited amount of daylight did not help matters any and it was very depressing, although most of us seemed to cope fairly well. The regular get-togethers with other couples helped out a lot. A low limit poker game was the activity most of the time, usually followed by a well prepared lunch ending the evening. The Coynes loved to play poker so it was mostly held in their home. Can you imagine six or eight people and their children watching sitting around a table with most everyone smoking.

With the approach of spring everything else seemed to change as well. The huge amounts of snow slowly began to disappear. The birds from the south, including the stately pelicans, were returning. Daylight hours took a drastic jump. The garden spot which I had worked on in the fall now needed more attention. The top soil I had collected and hauled in was levelled properly and with a shovel I had it turned over, mixing with the some of the sand this was the base. Annabelle was excited about planting some flowers and I left a suitable area for her next to the wall of our house on the south side, next to the vegetable garden area I would be planting. Since leaving my parents home many years earlier I had not had the opportunity to get involved with a garden. This was Annabelle's first practice run on flowers which she was looking forward to. Little Tony meanwhile was growing like a weed. His colic attacks subsided, and if he had an occasional episode we were capable of handling it. With the long winter blues now well behind us it was a pleasure to get outdoors. The garden was planted and everything grew very rapidly due to the long summer days. It was very expensive to purchase any fresh vegetable at the local stores. The fresh lettuce, onions and radishes was a great treat. Looking around it was obvious we were one of only a few families who had invested time and effort in a garden plot.

With the boating transportation season opening up a large number of new crews moved in. The closing down of the trapping season saw many families return to their summer homes in Smith. In general there was much more activity in the area once the winter ended. Like in many northern towns, community softball games were a great source of entertainment. There were some talented players on staff at the airport, so we were able to organize a team of our own. There was at least six other teams to compete with. Two teams, the RCMP and Northern Transportation fellows always gave us a run for our money with us often losing games to them. I was no great star when it came to playing ball but I took on a position as right fielder.

During this season work at the station increased with much more traffic in the skies above Fort Smith. With annual leave approaching already, I had to spend more time carrying out office duties in preparation of leaving. Speaking of annual leaves, every July Kevin, our officer in charge, booked his annual leave for the month. He owned a cabin in Ma-Me-O, a scenic spot thirty miles south of Edmonton. The family would all spend their leisure time there swimming, fishing and generally relaxing. I was second in command and honestly not looking forward to the additional responsibilities. After taking on the extra duties of the army signal corp., including commercial traffic it meant a great deal more work to handle. Kevin was very good in briefing me and had already provided extra training. Between the two stations we had a great deal of radio equipment to maintain. We did not have a full time technical staff, but were trained to handle minor problems. In the event of a major shut down or malfunction a qualified tech from Edmonton would have to be flown in. Due to the high level of noise and radio static interference, the majority of our transmitters and receivers were set up in a building three miles from our control office, handled by a remote wire system. The cable wiring was in a lead casing and mounted on poles, similar to the telephone system. There were a couple dozen sets of wires encased in this large cable. Before Kevin left, all of this system was in good working order; or so it seemed. On July 2 Kevin, his wife Lil and their ten children all boarded a CPA DC3 bound for Edmonton. I was now the man in charge, and things were going very smoothly. All staff members were very cooperative in assisting me with whatever duties needed to be taken care of.

Equipment Malfunction

As luck would have it, the easy part would soon come to an end. About a week and a half after the Coyne family departed, our office started to encounter period of intermittent operations of some of our equipment. From what we could observe, the problem was in the remote control system. The system continued to break down, shutting down additional equipment. All of our emergency equipment which was required to handle air traffic still operated since we were able to switch over from remote to base operation. The equipment most affected was the handling of C.W. traffic at the airport. The cause of our equipment malfunction however, remained a mystery. During this period we were encountering steady periods of rain. Finally someone informed us that one problem rain occasionally caused was breaks or cracks in the lead encased cable. If there was a break and water got through to the paper covered control lines, it could cause lines to short out. We had to assume the problem was in the lead cable. Max Hamp came over to our department from the army signal corp. He had some basic knowledge about the cable system that had been turned over to us by the army. There was a good three miles of cable to go over to try and locate where a break may have occurred. We did have a couple of meters that could identify shorts but first we would walk the line to visually view possible damage by a fallen tree or similar object. Given the severity of our problem, techs from Edmonton were on standby to come to Smith; but first we would make an effort to locate the fault. It continued to rain cats and dogs so to say the least, it was not very pleasant to carry out our investigation. Some of the staff was kind enough to help out. Using our meters we were able to establish there was a short in the line, but locating it was like trying to find a needle in the hay stack. The final area we pinpointed could be anywhere within a distance of a couple hundred yards. The other problem we had to contend with was the cable was strung about twelve feet above ground and strung between poles. We suspected the most likely place where a break in the cable could have occurred would be near the strapping by the poles. We spent many long hours in the rain trying to sort out this mess. To make a long story short, we did finally locate the problem. A small hole, no larger than the size of a fifty cent piece, was found in the lead casing. As suspected, it was near the pole and on further investigation chew marks were found in the lead cable covering. The only conclusion that could be reached was that it was caused by a squirrel, many of which living in the vicinity. We were overjoyed with the discovery immediately advising Edmonton Central office that tech services would not be required, at the same time requesting the services of a cable specialist to do the necessary repairs. A few days later all operations resumed as usual. I was one happy camper to see Kevin return from his annual leave at the end of July. He jokingly mentioned this couldn't have happened to a nicer fellow, meaning me.

The summer seemed to pass quickly between my working overtime and any spare time that we were able to find spent fishing or camping. Usually we were in the company of the Bourgets. Salt River was a favourite spot. Our RCMP friend Dave Hawker became engaged to the blond McDonald gal from the post office, followed by a quick wedding- a low key affair attended only by a few friends and family. We were among them. The big problem Dave encountered was that he was required to end his term with the mounted police as a result. For this idiotic law Dave had to give up a career he loved and was excellent at. Dave did not linger very long after his dismissal; he was quickly hired by the Federal fisheries department as an inspector. He and his new bride moved to Hay River N.W.T shortly after.

Talking to the family I was told that a cousin of mine had expressed a great amount of interest in what I was doing for a career. John Shepert, a first cousin, had decided to attend the Radio College in Calgary. He acquired the necessary qualifications and was hired shortly after graduation. John was brought up in the small town of St. Paul, and he too was interested in transferring to a northern location.

Nicholas Verbisky

To my surprise John contacted me via C.W. radio, sharing with me news of his posting to Embarras Portage. I knew there would be many things I would be able to advise him on in that location and marvelled at the coincidence of it all.

Our Return to Fort McMurray

Through the grapevine of our communication systems it came to my attention that Curt Hart, O.I.C. of Fort McMurray, had accepted a position as a technician and would soon be transferred to Edmonton. His new duties would consist of traveling to stations out of the Edmonton region including all the northern sites. If there was a major malfunction of equipment or if new installation was required, it would be his job to carry out the necessary repairs. Based on my seniority in the department and with my credentials I felt I had a chance in qualifying for the OIC position in Fort McMurray. Annabelle and I discussed the good possibility of returning to Fort McMurray if I applied and secured the OIC position. She was in favour of the move, and would happily return back to her old stomping grounds. The position, if I got it, would be a great benefit to me as well. There would be a considerable increase in wages and I would no longer be required to work shifts. It was only a couple of weeks later we received notification that the position of officer in charge and applications were being accepted. I discussed it with Kevin and he encouraged me to apply. Although he appreciated the extra help he knew I would be a good candidate and he also assured me he would give me a good rating on my application. I expected to be waiting at least a month or so before a final decision would be reached.

O.I.C. Appointment

In the meantime daily life carried on as usual. I decided I would feel okay if someone more qualified ended up with the position. If however I did secure it we would be able to move there without too many problems. The car was one thing we would be sad to leave behind. Our 1951 Plymouth was still our pride and joy. Hans Fruh, one of the operators, really took a liking to this vehicle and offered to buy it at a good price. A month and a half after my application was submitted the news finally came. Yes, my application qualified me for the position of officer in charge at Fort McMurray. I promptly accepted. After spending nearly two years at Fort Smith I was looking forward to the new challenge. Headquarters recommended that the move should be made at the earliest possible date. We were being sent out in style with a well attended farewell party, held in the lounge area of the single men's quarters. There were definitely some very mixed feelings leaving our northern home; we had gained some good friends in Fort Smith. Nevertheless, we said our goodbyes, and boarded a CPA DC3 flight into Fort McMurray. Our personal effects would be shipped a few days later on CPA C46 cargo carrier. As we got off our flight in Fort McMurray we were met by Curt Hart and a few other old friends. Curt told us his family was ready to relocate to Edmonton. He would however be spending a week or so briefing me on the operation. One major change I had to contend with was taking over the Army signal Station. Other than that there had been few changes since my transfer to Fort Smith. The army signal set-up was located in the town- there was a wireless station, two single unit married quarters and two duplex units and a double garage. This was all located in one large compound only a block from the main street which was the centre of town. I agreed to move into the single unit in town which located next door to the wireless station. Dropping in to visit the airport operator office I noticed few changes in the general set up. One thing I did note was a considerable change in staff. Fred Burwell and Ed Burrows were still around, but Ed was moving to Hay River where he was offered the O.I.C. position. Most fellows had been posted here after my move to Smith. Pete Wellman recently moved from Embarras Portage. He of course had married Edna Letourneau, a very close friend during my work in Embarras. Next door to the Wellman's were Andy Anderson and his family. Andy was an ex-army signal member who had been station there when they disbanded their operation in McMurray, and simply had transferred over. We were very excited about moving into our new home. It was an older dwelling, still in very nice condition though and included a large fenced yard. Our next door neighbours were Billy and Shirley Hill. We knew them really well from our previous lives in Fort McMurray. Billy and his parents owned Hills Drugs, a very old establishment in town. The Hills had two small children, one who was Tonys age. The other single unit was occupied by Roy Hawkins, his wife and three children. After returning from overseas, Roy was employed with Canadian Pacific Airlines as an agent. While being in the army as an intelligence officer, Roy learned and used international code in CW radio communications. He studied and passed the necessary exams qualifying him for employment with our Department. Roy also had considerable training in transmitting and receiving equipment which made him a good candidate to assist me with the maintenance of the station. While becoming adjusted to my new role Curt was a good trainer and provided me everything I'd need to know for running the Aeradio office at the airport as well as the downtown ex-army signal station smoothly. The army signal stations were originally operated by a six member staff. As we took over a large amount of the commercial traffic we were able to downsize the town office to three staff. There was a considerable amount of space in the ex-signal office in town so I set my office up there and traveled to the airport only when required. The government was very generous in providing me with a new Austin four wheel drive jeep. This unit was also used for all airport staff shift changes. After a couple of weeks of orientation Curt and his family departed for Edmonton with mixed feelings. There was no question they had enjoyed the low key lifestyle Fort McMurray offered, and Curt loved his job. However being offered a promotion with an increase in wages, it was an offer that was hard to refuse. The couple were also looking forward to the better school system the city could offer their children.

Once I took over the workload in our department seemed to substantially increase. One of the major reasons for aircraft arrivals was due to the building of radar sites in the arctic and mid-Canada line. The mid-Canada line had its' main site developed in Stony Mountain twenty five miles south near the hamlet of Anzac. The main site in the arctic was Cambridge Bay N.W.T. Aircrafts destined for Cambridge Bay would leave Edmonton fully loaded with cargo. The mostly DC4's and British Bristols would than land in Fort McMurray to top off their fuel loads before proceeding north. Cargo for Stony Mountain was brought into McMurray then forwarded over to the mountain by air force choppers.

As we settled into the new home we enjoyed renewed friendship with many of the people with whom we had previously visited. One of the married couple we had a very close relationship with was Steve and Shirley Brooks who were glad to see us come back. Steve worked for the local power company as a diesel engineer. He was one of the senior operators at the power house which generated electricity for the Fort McMurray and Waterways. It was Steve who tipped me off to a potential money making deal that I simply couldn't turn down.

Chicken Farm Operation

Steve was born and raised in Fort McMurray and very knowledgeable about what was happening around town. He told me one day that a number of lots were going to be sold for taxes. They were on the north end of town, above the flood level area.

I had always had an interest in operating a poultry farm with egg laying production, although until that time it was more of a hobby. It was a known fact that the Fort McMurray area had never seen a chicken farm operation. By owning a farm in the Vilna area I assumed getting the necessary feed shipped in shouldn't be a problem, and decided it would be neat to give my dream a shot. I made contact with the town office and they did have some lots grouped together for sale, a couple of blocks north-east of our home. The lots were covered with a growth of trees on level ground. I was initially very excited about the prospect of this little side business, and did not hesitate to purchase three of the lots for a total price of $65. Thinking back on this, knowing the value of the land in Fort McMurray in 2004; it makes me wish I had held on to these properties. I promptly put my plans into motions. I'd done some research into my plan: conducted a feasibility study on poultry production using information from the U of A animal science department. There were a number of issues that needed to be considered if I was to be successful. The type of hens that would be the best layers, the building required for a brooder house, type of brooder equipment as well as a suitable building for egg production were all questions I needed answers for. Shipping in the chicken feed via Lac La Biche had to be discussed with Dad. This would no doubt require a considerable amount of extra work on his part. It was also obvious this operation would cost me a considerable amount of money before the first egg was laid at least six months down the road. I moved forward patiently with my plans one step at a time. The first job was to have the lots cleared and leveled so I secured the services of a local cat operator and the job was done quickly for a couple of hundred bucks. Next I located a rundown ten by twelve feet building which was solidly built but needed a new roof. I got this for a hundred dollars and had it moved on the property. In the Eaton's Catalogue I'd seen a chick brooder with capacity for up to six hundred chicks. My plan was to have at least six hundred baby hens so I ordered those too. My research told me the most profitable egg layers found were triple A Leghorns. What I needed the most still was a building that would adequately house six hundred adult triple A hens. As luck would be, I did soon locate such a building. It was a surplus pre-fab unit of four by ten panels, fully insulated against the elements, twenty by forty in size. The panels included doors from both ends and a number of enclosed windows. The building was supported on 8 x 8 inch twin beams. I was told that it would take only a couple of days to completely assemble it with two or three people helping. The total cost was only five hundred dollars, which I thought was an excellent deal. I was extremely pleased in the way my business plan was falling into place. I got a hold of Dad about the grain deliveries. He felt that the whole operation was going to be a large undertaking, but assured me he would be able to handle his part of the bargain.

With all of the planning taking place for the egg farm, we also had some very good news; there was a new addition coming to our family. Our second child was due to arrive sometime in June. Little Anthony would have a new brother or sister. Annabelle was in great shape, feeling good, putting on a few extra pounds. Spring was approaching so I had to quickly wrap up some final steps for the chicken farm. My brooder had arrived from the Eaton's catalogue so I had the brooder house disinfected, cleaned and insulated in anticipation for its new residents. I placed an order with a chick hatchery in Edmonton for six hundred baby leghorns. They were shipped in by CPA air and arrived on April 29, 1960. Quickly I had them transported to the brooder house where the temperature was maintained at 95 degrees F. This warm temperature would be maintained for two weeks, then reduced a couple of degrees a week until the chicks were six weeks old and ready to move to the still unassembled new house. While this was taking place, Steve Brooks and another chap helped me put together the prefab. The power company had a pole put in and some basic electrical installation made. After six weeks in the brooder and with the chicks growing rapidly which caused some crowding in the brooder. With good babysitting and constant care I only lost four of the little birds. In the prefab the chicks had a lot of space. There was a small oil heater installed in one corner. The brooder house was converted into a feed and storage building. I ordered thirty sacks of chop (mixture of oats and wheat) as well as some whole oats for my hungry investment, which Dad soon delivered. This every day event was costing a considerable amount of time and money. Even with everything running smoothly, I would not see any returns until at least October. The little critters were growing rapidly, with daily care and attention necessary. The calendar drew closer to Annabelles due date, and I would soon have another little one to worry about as well.

Pamelia Arrives

Ann had been seeing Dr. Pickford during the pregnancy and he told her the baby was due in early June. As predicted, a baby girl arrived on June 13, 1960. Annabelle had an easy delivery with the doctor and nursing sisters in the small Fort McMurray hospital providing her with loads of TLC. Mom and baby came home a week later. If it was a girl we had picked a name of Pamelia, which was the name of Annabelle's grandmother. Tony was very excited with having a little sister; we all were very excited. The first month passed quickly. July was soon upon us, with the summer weather in the north being hot and pleasant. Annabelle's parents and her sister Helen were planning on coming to visit with us and see little Pam. They made the trip from Powell River as far as Lac La Biche with their car. From there they hopped on the NAR muskeg express passenger train. The visit was exciting for them in a lot of ways. It was the Adami's first return to Fort McMurray after being away for quite a few years and a chance to get reacquainted with many of their old friends.

The hot July weather was a perfect time to go out on picnics and enjoy the otherwise desolate scenery around McMurray. Many of these outdoor lunches were enjoyed at the wing dam of the syne of the Clearwater River. Every Sunday there would be a great gathering of families at the site. Fires would be lit for wiener roasts and almost everyone went swimming. It was the location of my first experience with water skiing. Steve Brooks was an avid water skier. He talked me into trying it out. I was definitely leery but being young at heart and a bit of a daredevil to boot so I gave it a try. Luckily, nothing went wrong on my first attempt but I do recall a number of spectacular wipe outs from behind Steve's boat, much to the amusement of everyone enjoying their lunch on the shore. Billy Hill, our next door neighbour was usually on the water. Billy owned Hills Drugs with his dad Walter. Both were druggists by trade. His Dad and Mom came to Fort McMurray in the forties. Their drug store was very small but served the community well. They even sold insurance out of the tiny building. Pete Krawhuk was another avid skier. He was a partner in the local Fort McMurray hotel. Steve, Bill and Peter all owned boats, nothing fancy with 25 and 35 hp motors. Each boat could only pull one skier at a time. I clearly recall my first attempt on getting up on skies. I had the tendency to hold on to the rope which would cause one to be pulled down underwater. It would mean swallowing a great deal of water. I soon learned to release upon falling. Skiing on the clear river water was also very tricky. There were numerous currents which would cause the skies to vibrate. Additionally there were also snags in the water which had to be avoided. Northern lake water skiing is not for the faint at heart! It was just one of the activities the Adami's enjoyed during their time with us that summer.

Occasionally my mind would drift back to the time I was stationed in Smith River Airbase. Often the memory of Jim Bresdon would enter the picture. Jim was not only my supervisor, but a mentor and close friend. He strongly encouraged and helped me to study for my aeradio barrier exam. He suggested that to take it would improve my ability for promotion, and increase my rate of pay and he was right. Jim was very gifted and knowledgeable in aircraft instrument landing systems. He was recruited into the Edmonton Regional Office as a Technical Specialist. Being a commercial pilot he was very suitable for this job. Jim also attended an Aeronautical Training Center for the necessary orientation. While stationed in Edmonton Jim visited me on a couple of occasions in Ft. McMurray. Each time flying in with his beloved Cessna CF-FCR. because of his exceptional ability, he became well known through the region. Jim's expertise was soon learned by a large city in South America which required updating of their interment landing systems. This lead to a request for a loan from the D.O.T for his services. Jim agreed to this challenge and soon after proceeded on the long journey with his own Cessna. He felt his airplane was well equipped to conclude this long journey, being it possessed all proper updated instrumentation and long range tanks. After arriving safely it is known that Jim spent approximately two months on the assignment. His return trip took him over the rainforest including of long distances of unforgiving and desolate terrain. Somewhere across that land mass he disappeared. Exhaustive searches proved fruitless. No one ever found any signs of Jim or his aircraft. Jim is one person I will always sadly remember as being a true friend who perished too soon.

Family Visit

Oscar, Mary and Helen spent a busy two weeks with us the summer Pamelia was born. Helen was a busy auntie, playing with Tony and little Pamelia. She was also very intrigued with the leghorn chicks which by now were nearly grown. The in-laws spent a considerable amount of their time exchanging experiences with Ann about their new homes. It was obvious they missed their McMurray friends, having spent some fun years in the area. Their departure for home was a sad one. We all knew it would be some time before we would be getting together again. It was 1960, and we all lived a very busy life. Annabelle had her hands full looking after Tony and Pam. I was pretty well tied up both at work and babysitting my chicks. There was the occasional crisis that required more time and attention including staff changes, illness, but the most common problem being radio equipment failure. It was imperative that all radio communication in the aviation sector was of the highest degree. Frequent maintenance was ongoing. Roy Hawkins was one of my aeradio staff, but because of his ability in radio maintenance some of his time was allotted to that area. The numerous radio receivers and transmitters were checked over and adjusted at least once per month. A good portion of the work would involve checking and replacing tubes. We were also required to service and maintain a remotely located emergency power plant. This unit would cut in automatically to produce power for our equipment in the event of a commercial power failure.

I had a very unusual occurrence with one of my staff about this time. Don Ross, his wife and three small boys lived at one of the duplex units in the town. Generally Don stayed very close to home and was very protective of his family. Little by little we started to notice some strange behaviour in the otherwise normal fellow. It was noted first by the other operators who had to share shifts with him. He began to be paranoid about aircraft flying overhead, and suspected they would be dropping bombs. He also began to have difficulty communicating with these aircraft as well. Once this problem became quite pronounced I immediately removed him of duty and assigned him to medical leave. Obviously, there had been some kind of a breakdown in his mental condition. One day he was also found driving the seven miles to the airport with the distinct feeling he would be attacked by someone. There was very little medical help for people with mental conditions in Fort McMurray. After a short leave, with rest and medication Don did improve somewhat and claimed he was able to go back to work. That did not last very long, and he again went into a state of depression. Regional Office suggested after medical leave he should return to his home province of B.C. where he had relatives. Don submitted his resignation with no plans to return back to Fort McMurray. We never did find out exactly what the issue with him was.

In mid August my sister Ollie and brother in law Ed paid us a surprise visit by flying in with CP Air. They came in for a week, spending part of their holidays in our home. Ollie and Ed enjoyed the peaceful life style of the northern community. We entertained them in the best way we knew by introducing them to our friends, picnics at the wing-dam and I helped with displaying my "talent" for water skiing on the Clearwater River.

Roy Hawkins, one of my sidekicks lived in one of the government homes just down the street from us. Roy was a long time resident of Fort McMurray, very involved in the town's affairs. He was fire chief and asked me to join on as a fireman. I agreed to join the group. There was about half a dozen local young fellows who were very active in this department. The equipment was very primitive by current standards but suitable in those days. Major equipment consisted of three ton truck with a five hundred water tank on the rear was amounted water pump operated by an air-cool motor. There was a second truck also with a large water tank which acted as a water supply unit. There were no buildings in the town higher than two stories', so with these two trucks we could handle most fires effectively. We usually had fire drills every week and all fire men were volunteers. Everyone was provided with proper fire gear such as coats, boots, mitts and hats. Occasionally after a fire practice the fellows would to the local hotel for a beer so it was a social activity as much as anything, an important part of the community spirit. I was involved in quite a few fires; some homes we saved, others we did not.

With fall ahead, my hens were maturing and that meant egg production- and the returns on my investment- should be close at hand. In my limited spare time I constructed a couple dozen egg laying nests, mounted horizontally about two feet above the floor against one wall. From a nearby sawmill I got a supply of shavings which I used as liners and to cover the floor. In part of the building I erected roosts constructed of 2 x 2 materials with dropping pits below, screened off with fox wire. As the chickens would be requiring a constant supply of clean water I had two 45 gallon barrels welded one on top of each other. I hooked an automatic water system to these with water supply brought in weekly by a water truck operator who also provided water needs to most residents in town. I cleaned the building once a week and spread fresh litter. The first pullet egg arrived on September 4, 1960. This was highlighted as a very special date in the life of Silver Shell Poultry farms operation. Every day we seemed to have a new surprise at Silver Shell. First there were only a couple of these tiny eggs, but as the days went by more and more appeared in the nests. Annabelle was impressed; at least we did not have to buy any more eggs but it did take nearly a dozen to provide a good meal for the family. There were also some cracked eggs that we gave away to friends free of charge. Circumstances changed drastically about a month after the first small eggs arrived. The size of the eggs increased to normal and from that point on production was almost overwhelming, with about fifteen to twenty dozen eggs a day to contend with. Egg collection was a daily task which required care and attention. Some eggs were slightly soiled even though the nests were kept very clean and Annabelle helped a great deal with this task and even little Tony took an interest, although his help wasn't always so helpful with him thinking they were toys to play with. From Edmonton I ordered five hundred one dozen cartons to which we affixed the Silver Shell Poultry Farm stamp. Everybody in town learned about the fresh eggs we were selling and it got to a stage we could not keep up with the demand at the price of one dollar a dozen. I noticed in the Eaton's catalogue egg washers for sale. I ordered one of these units for thirty five dollars. This egg washer was a simple unit made up of two wire egg baskets each holding about ten dozen eggs. The wire egg basket when full of eggs was lowered into a pail like tank filled with warm water and soapy solution was poured in. A small electric motor gently rocked the whole unit slightly in one direction then in another. The eggs in the basket were suspended slightly and then gentle rubbing would clean them in about 10 minutes. They were placed in cartons overnight, now ready for sale. The cleaning of the eggs for sale was always done after hours and on weekends, which left us very little time that wasn't spent working either at the farm or the station, or at home for that matter, but we were alright with that.

Working a straight day shift was one of the perks of being an officer in charge. My office was very small, located next door to our operations centre. With only nine staff it was less of a problem then in Fort Smith when it came down to drawing up the weekly shift plans. Every day there was some type of correspondence or other paperwork to contend with. There were extra considerations with the staff as well, relating to annual leave, sick leave and overtime; not to mention reports on the overall operation of equipment. That included living quarters. The department provided staff with six married units. These homes were fully furnished, and inventory had to be maintained. If and when some items were broken beyond repair, they were written off and new ones ordered. Each one of us in married quarters were required to pay rent of forty five dollars a month which included utilities. During the winter months the oil bill alone usually exceeded that, so we were getting a bargain there.

Aircraft Malfunction

One morning while I was busy with paperwork in my office, I heard Fred Burwell communicating with an aircraft. It was a DC4, landing to top up fuel enroute to Cambridge Bay N.W.T with a full load of cargo for the radar site. After fueling up the large plane taxied to the runways end for take-off. Suddenly there was a great roar and very unusually high pitched noises coming from the runway. It appeared the pilot had aborted take-off for some reason. The aircraft had come to a stop past the midway down the runway. Immediately the fire truck was alerted and we all quickly made our way to the aircraft on foot to see what had happened. We noticed the nose of the aircraft was physically on the runway. The pilot and co-pilot had left the aircraft and were taking a survey of the damage. The propellers of two of the four motors had made contact. The blades had been sheared off and pieces flung in many directions, a couple even embedded in the aircraft- some very close to the air crews cabin. They walked around, shaking their heads and thanking their lucky stars for not receiving any injuries. The cargo was secure, but caused some shifting in the aircraft due to the sudden stop. The pilot and co-pilot could not come up with an exact reason for the malfunction- although they did know why it had occurred. The aircraft didn't have sufficient speed for take-off causing it to tilt forward. They would not make comment as to whether this was pilot error, or an electrical malfunction but their silence suggested the former. The departmental airport crew was able to move the aircraft off the runway temporarily. The accident was reported by me to the Department of Transport (DOT) inspection branch. A crew arrived the following day to investigate. A couple weeks later, the motors were replaced and the plane was ready for take- off. They were again north bound with their cargo. This time, the plane made it off the runway with no complications.

About this time I was needing a personal vehicle. Although I had full use of a government jeep, with starting up the chicken farm I needed something for hauling feed, equipment and other supplies. I was lucky in locating a half ton pickup owned by Lawrence Tolen, a fellow who operated a water delivery business in town. It was a '56 Ford, in nice condition. Due to the lack of roads in the area it had very low mileage. Steve Kowach, a mechanic at the airport had it tested out and stated it being in excellent mechanical condition. Apart from few minor dints and paint scratches it was perfect, and the price was right at six hundred dollars. Annabelle was happy to have a vehicle to get around as well.

I had gotten to know a fellow by the name of John A. McDonald. He lived in a small shack near my chicken barn and would come over frequently while I was tending to the flock and chat. He talked openly about his lifelong experience as a resident of Fort McMurray. John was pushing sixty, a tall distinguished looking gentlemen of Métis decent. He claimed he had working on the river boots since the age of fourteen. He ended up as a Captain with the Northern Transportation Company. Apparently he got released from his duties a few years earlier, and although he didn't say the exact reason, I suspected he must have had a drinking problem. I liked him however, and obviously his finances were limited. I started giving him some of the slightly cracked eggs which weren't suitable for sale. In return John would sometimes help me with minor chores and since he lived close by, he also agreed to keep an eye on the property. Annabelle also took a liking to this fellow, felt sorry for him. He claimed he didn't have any friends and spoke of his family. His wife developed a mental problem and was admitted to an institution near Edmonton. He had three adult children but had very little contact with them. Sometimes he would be invited into our home for conversation and supper. He was always very thankful for the meals and company. John played a fiddle and brought it to our home which was very entertaining and enjoyed by little Tony and Pam.

Egg production was booming, but I had found a pamphlet which suggested I could increase egg laying capacity even more. What I needed to do was increase the daylight hours with lights both in morning and evening. I found a timer, set it to come on at 6 am and again kicked in at 5 pm. If this was not done the layers would be spending more time on the roosts not laying. Even though things were going well, another major problem reared its ugly head: signs of cannibalism. Perhaps due to sheer boredom, the layers started pecking at each others hinds causing bleeding. One suggestion to stop this nasty habit was to apply pine tar to the pecked areas which would leave an unfavourable taste in the beaks to discourage any further pecking. This worked for a short time. It was not the cure however and further pecking always occurred. It became so severe that sometime the whole hind end of the chickens was pulled out, causing death. With the loss of a number of these layers, and at a loss myself, I had to consider some extreme measures. I contact the U of A Department of Animal science poultry division for advice on this matter. They suggested each bird is fitted with plastic glasses called "specs". These plastic shades could be mounted on the beck with a cutter pin threw the nostril section. By having this done the bird is able to see to the sides and down but cannot see directly ahead. She would be unable to peck at the bird directly ahead of her. I ordered six hundred specs through the T. Eaton's catalogue. Our good friend Ernie Duchaine who lived a couple of doors down had agreed to help me with the tedious task of securing the specs on each of the chickens. The first couple hundred chickens were fairly easy, but trying to catch more of them after that was overwhelming. Some comedic relief was provided to see them stagger around after the spec had been placed on. When they tried to peck they would miss their target. It was an exhausting two day job for both of us but the pecking disappeared.

For the next couple of months everything seemed to roll along smoothly. The egg laying business was profitable, but there was a considerable amount of work. Luckily I had Annabelle to assist with the egg picking, cleaning and packing in cartons and ready for sale. Dad and my brother Frank helped by having the feed shipped in feed for me regularly. The flock required a one hundred pound sack per day, mixed with the laying mesh. This feed was crushed with a combination of oats and barley. They also had a good supply of oyster shells available which was needed for the normal development of the egg shells. By carefully watching the flock it was important to remove any sick hens or poor producers. Culling reduced the amount of feed required to produce a dozen eggs thus improving efficiency which was even more important in this remote location. With the cold weather now upon us it, it required extra care and attention to keep the temperature in the barn comfortable. I'd installed small oil heater in one corner which was enclosed with a wire cage like fence. An even temperature was maintained with an exhaust fan. The manure was removed once a week, fresh shavings were applied to the open floor area and nests. There was always plenty of manure on hand and a few people in town came and took a load to apply on the garden areas. I have to say I had some anger displayed toward me…the cured manure produced such an obnoxious smell that the pungent, unmistakable odour could be detected pretty well throughout the whole town.

Winter was again descending in Fort McMurray and attacked us with another severe blow. Snow was falling almost daily and cold temperatures plummeted downward. We all settled in for what appeared to be a long haul. With Christmas around the corner it was beginning to be an exciting time. Pam was too little to understand, but Tony was told about Santa who would bring him presents. I remember as a small boy myself I would look through the Eaton's catalogue admiring all the different toys. One thing that always stood in my mind was paddling cars. I would dream of some day owning one and being able to paddle it all over the country. Annabelle and I noticed a similar one in the current Eaton's catalogue and decided to order it for Tony. It was an army jeep, made of heavy metal and was blue in color and affixed with army decals; and it arrived disassembled. Just before Christmas our good friend Ernie Duchaine helped me to assemble this prize toy which was then put away for safekeeping in the shed. Our annual Christmas party was planned for at the airport, as usual in the lounge area of the single men's barracks. Bill Mitchell, "Uncle Bill" as Annabelle knew him, was the cook and he had prepared an excellent meal attended by nearly everyone associated with the Department at the airport. Tony, Pam and other young children were excited to see Santa was there handing out small gifts. Even though we had children of our own, Ann and I both missed our families at this very special time of the year.

Brother Stan had written us a letter saying he would be coming to visit us for a few days over the holidays which was a welcomed surprise for us. Stan arrived via CPA two days before Christmas day. We all attended mass on Christmas Eve and after getting home Tony and Pam went to sleep without any difficulty. After having a few drinks and snacks with the Duchaines we got busy setting out Christmas presents. Stan was very generous by bringing a number of toys and not having children of his own, he had done a fine job in picking up some fun items. We were most excited though about the paddling jeep for Tony. Morning came early with a great deal of excitement around the Christmas tree. Pam, being a small baby, did not express any great enthusiasm. Tony attacked the wrapped presents with a joyful expression. When it came to the Jeep he took a few looks at it, then carried on playing with the other toys he'd received.

Later that day we had planned a lovely Christmas dinner, including a turkey that mom had sent with Stan. We always made a point of inviting some bachelor who was alone at Christmas. This year we asked the airport manager Lloyd Goltz to join us for our festive dinner. Lloyd arrived late, and four sheets to the wind. Due to his condition suffice to say it was not the most pleasant meal. We made the best of it before sending him off to the bedroom to sleep it off. When he awoke he was most apologetic for his behaviour. Stan went home a few days before the New Year. He told us he found the Fort McMurray area interesting, and I was glad we had made a point of taking him to see the few highlights of our small community during his visit. There was a New Year's party held at the airport which we all attended.

Plans to Leave the North

The town was quiet after the New Year, the weather remaining cold and severe. During the long early months of the year, inside our home away from the unforgiving elements, Annabelle and I discussed leaving Fort McMurray and moving somewhere less isolated. We'd spent years in this restricted environment and were ready for a place which would give us a better access to the outside world. There was no question we would find a better lifestyle in a southerly location, but there were other factors to consider. I would be required to give up my officer in charge position, and the loss of our northern allowance pay was no small dice. There was also the issue of my returning to working shift work, which did not really appeal to me. I *was* ready to give up the chicken farm operation. When the layers are two years of age, their egg production had decreased to a state that is no longer profitable, so to continue would mean replacing the flock with new chicks in the spring. Even though I enjoyed the challenge and it was a profitable venture, it also required a great deal of hard work and dedication without break. My greatest challenge was having Dad prepare and ship me chicken feed on a monthly basis. I knew it was not easy for him, having to drive to Lac La Biche to catch the NAR freight for the north. We made up our minds to close down our poultry farm operations come spring. After weighing the pros and cons we had to decided that I would be requesting a transfer to a more southern station. I wrote a memo to the Regional Director outlining a request for a southern transfer. Ann and I agreed to suggest Lac La Biche as a location we agreed on, although the choice mostly favoured me because my parent's home was close at hand- however Annabelle liked the idea as well. I was a faithful servant in the Department and was fairly certain I would soon be receiving a favourable reply. The response to my memo stated that my name would be entered in the event a vacancy occured in Lac La Biche.

Annabelle and myself with our first car

OIC at Ft. McMurray with Govt Jeep

RECEIPT (Taxes)
TOWN OF McMURRAY
Province of Alberta

Nº 4467

McMurray, Alberta, *March 1st* 19*60*

ON ACCOUNT OF *Sale Of Lots* $ *65.00*

THE SUM OF *Sixty Five* —————— $\frac{00}{100}$ DOLLARS

RECEIVED FROM *N. Verbisky*

Cash Book Folio No _____

Account Number	Description of Land			Amalgamated Taxes					TOTAL
	Lot	Blk.	Plan Number	Current	Discount	Arrears			
	19	4	616						
	20	4	616						
	21	4	616						65 00
			TOTALS						
RECEIPT (General Other Than Taxes)									

PAID BY: Cheque _____ Money Order _____ Cash ✓ _____ Labour _____

TOTAL 65 00

Signed *Joan Reebee*
Secretary-Treasurer

Municipal Supplies Ltd., Edmonton & Calgary

Fort McMurray lot taxes

aeradio office letterhead

Fort McMurray Airport, Christmas card 1956

Airport staff at a family birthday Ft. Smith N.W.T.

R.C.M.P dog team at Ft. Smith N.W.T.

Miss Annabelle Adami
Box 649,
Powell River,
B.C.

Mr. Nick Therbiaky
℅ Radia D.O.T.
McMurray
Alberta

Letters

SILVER SHELL
POULTRY FARM
McMURRAY · ALTA.

DEPARTMENT OF TRANSPORT
CANADA
54

ANN'S CONFECTIONERY
LAC LA BICHE, — ALBERTA

Our business letterhead

Our first rented home

Annabelles first family home on the trapline on the Mac Millian

Army Signal Station

Fort McMurray airservice Norsemen on weekly mail and supply run

Chicken Farm Closure

In the meantime, closing down the chicken farm presented some obstacles, the main one being getting rid of nearly six hundred two year old chickens. Raising these birds from chicks I had grown attached to the birds, and they felt more like pets than property. I made it known in town that I was discontinuing the egg laying operation and would sell the whole set up as is. Not surprisingly, there were no takers so I had no choice but to proceed with the slaughter. I decided on a time, the later part of February and month of March. Hopefully by then, the weather would improve and would make the whole unhappy project a little easier. I told dad when the last shipment of feed should occur and he was pleased. I made posters and had them distributed around town. "Chickens for sale" one dollar each, live or otherwise. Orders came quickly. I was hoping more people would buy live, which was a better deal for me. The dreadful day finally arrived, and I had tears in my eyes as I made one of my final trips to the barn to begin the unpleasant task. People would arrive, and take a dozen or so live. In those cases I would give them an extra bird free of charge. Getting rid of the first hundred or so was the most difficult. After that I kind of got used to the chopping block.

The most unusual order came from two forestry fellows from the Anzac who drove up in a Volkswagen car. They were four sheets to the wind after spending a couple of hours at the local hotel bar. They wanted thirty live chickens thrown inside. I was surprised, and began to toss the wildly flapping birds into the small vehicle. Needless to say, the fellows had some difficulty getting back into their seats. As I watched them drive away, I could only imagine what it would be like to drive thirty miles through a bush road back to Anzac with these birds flying around in the vehicle. Due to the age of the birds they were not tender, but suitable for being boiled or as an ingredient in stews. The majority of these layers were sold to natives who generally prepared them that way. The price was right and these leghorns were in very nice shape. By the end of March the flock was totally gone. My poultry operation was closed down, with some degree of sadness. Some items from the barns were sold others packed up to be shipped whenever my transfer would finally take place.

Father Brown

As devoted Catholics we attended mass regularly. One of the most unusual and special Catholics we got to know was our parish priest, Father Brown. He was an American from a rich family. After a brief stint in the air force he had received his calling to become a priest. His preference was to serve people in the north, so he joined the oblate order. Father Brown was a very handsome and outgoing fellow, a few years past his 40th birthday. We were pleased to have him baptize our daughter Pam. We soon learned that Father Brown was gifted in so many ways and a great asset to the people he served. Despite only being posted in Fort McMurray for a short time, the efforts from his time there still remain. One of the major tasks he completed was relocating the abandoned small log mission church to a site on the new church grounds. This humble structure was restored to its original state, where special masses such as Easter were held. He was instrumental in constructing a retreat for the nuns that was being secretly built, about half a mile south of the church campground in a very secluded location. These dedicated grey nuns faithfully served the people in the area in so many ways. They were the nurses in the Catholic run hospital, and teachers in the Catholic school. Father Brown also acted as administrator at the hospital. One of the things everybody noticed was his well matched dog team, which he took great pride in. After Fort McMurray he headed to various locations in the north, always building churches as he moved along.

Lac La Biche Posting

I did not expect on receiving news from the Regional Directors Office so soon, but a letter came informing me my transfer was approved and I would be posted to Lac La Biche the first week of May. We slowly packed household things we weren't using. We would be provided with a fully furnished dwelling, so we only had only our personal items to take into account. With a growing family there was still a considerable amount of packing to do. Central Office had also informed me that the senior operator, George Kubinski would take over acting officer in charge duties after my departure. I spend almost a week briefing him on his new assignment. Of course, once word got around about our transfer a farewell party was in the planning. As usual it was held in the lounge area of the single men's barracks. The send off was a very special event, attended by nearly everybody at the airport was well as our friends from the town. Once again this move was a difficult one as we looked back at all the good friends, and the pleasant times we enjoyed in Fort Mac.

We boarded the NAR Northern Alberta Railway muskeg express arriving in Lac La Biche on May 4, 1962. There was about four inches of fresh snow on the railway platform when we exited the train, fresh faced and looking forward to our families newest adventure. On the platform we were greeted by Oscar Kleven, the officer in charge. I had spoken with Oscar many times, but never met him personally. He was a huge man, well over six feet tall and weighing perhaps 250 lbs. Despite his size it was obvious he possessed a very gentle, easygoing disposition. He shook my hand warmly, saying he was glad to have me on staff, being one of the single fellows had recently left for an arctic posting. We were driven to our new home on the northwest end of town. Our new home would be a duplex unit in a small government compound. This consisted of two duplexes and a single unit. The two duplexes housed the four operators families, while the single unit was for Arnie Amble the airport manager. Our unit looked very comfortable and well maintained. Three bedrooms on the second floor, and a full unfinished basement. We briefly explored our new home than Oscar invited us to his home for lunch. We were greeted at the door by his wife Gerta, who like Oscar appeared to possess a similar personality. They had two young very blond girls of similar age to Tony and Pam. As we waited for lunch to be served, Oscar told us he had been with the department since shortly after the war. He was stationed in Snag, Yukon as a single operator and had gotten married when he was in Teslin. Oscar was the officer in charge in Teslin before coming to Lac La Biche two years ago. While Annabelle and the children visited with Gerta, Oscar drove me to the airport. It was about two miles from town with the aeradio central office centrally located near the well groomed paved runway. I was introduced to Marshall Pysar. His family occupied the duplex unit next to ours. This station was designated as a compulsory reporting point for all aircraft flying via the Blue 84 airway. The operators were also required to prepare hourly weather reports which were disseminated to the station via teletype circuit. CW Radio was not used extensively at this location. I had to admit it was very clear that working here would be much less challenging then some of my previous locations, and if anything appeared to be somewhat of a retirement spot. The only downfall was the fact shift work was again required which did not appeal to me. I would spend a day or two with Oscar getting a handle on daily routine before being assigned a shift of my own. Oscar did not work any shift work. He only worked two day shifts a week, the rest of his time dedicated to office correspondence and maintaining equipment. It didn't take us long to meet all the people that worked at the airport. Mostly a friendly bunch; mostly young like ourselves with small children. Marshall and his wife Lucy had a family of five and fostered two additional children as well. Ivan and Sylvia Pitts who lived in an adjacent duplex had a boy and girl both preschool age.

Shortly after our arrival the rain began to fall on almost a daily basis. Mom and Dad had plans to come and visit us, but the roads became difficult to travel on. All the roads in the area had no pavement, were very narrow, poorly maintained and in most areas lacking gravel. The distance to travel was only about fifty miles as the crow flies, but there were no direct road connections. Two accessible roads meant travel would take about one and half hours in one case and about three hours in the other. The latter road was in much better condition. We were unable to travel to my parent's home as we had no vehicle. The half ton pickup was sold before leaving Fort McMurray. Our plan was to purchase a new vehicle from a Chevy dealership in Vilna as I knew the manager well. He sold us a brand new Chevy for the asking price of $3200. After the rains subsided and road conditions improved my parents were able to come for a visit. I drove back with them to Vilna to pick up our new vehicle-what a beauty it was. This gave us the luxury to travel and visit with many of our relatives and friends that we had little contact for quite a number of years.

As time dragged on I once again got itchy feet about getting involved in something other than work at the airport. I explored the idea about operating on automatically controlled dry cleaning business and laundromat. That idea fizzled when somebody got ahead of me. The second venture I had in mind was operating a confectionary store, which would be quite demanding both timewise and financially, but Annabelle agreed it may be a good idea. She could head off the operation and I would fill in with help on my time off from work. With the confectionary proposal in the planning stage we agreed to enjoy the summer. There were a great number of lakes in the area so we frequently enjoyed camping and fishing in many of these excellent locations. Often we would travel to these spots with a group of friends we quickly got to know. Tony and Pam always had playmates during these outings and loved the idea of roasting wieners for their hot dogs. A few of the people in our group would bring along their boat with a few of us heading out to take in some excellent fishing. Most lakes at the time were full of fish of different variety's on which there were no limits. What a joy it was to have a good fish cook out, then still have a bunch of fish to take home with you.

Ann's Confectionary

There was already a confectionary in town and in preparation for opening our own I got to know the fellow that ran it. I told him about my plans to start a new business and he didn't have any objections. The town's bylaw stated that grocery stores closed at 6 p.m. during the week, as well as being closed every Sunday and Wednesday afternoon. It was obvious that the bulk of our business would be during the hours which the grocery stores were closed. I looked for a spot in town suitable for this venture. We preferred to have it across town from the other store. I came across a small store front area that was a portion of a bakery. It was in the centre of town and owned by Philip Patry. I approached him about my plan and he felt it was a good idea that would benefit his business. Annabelle and I carefully looked over this store front location which Mr. Patry had previously used for storage. It was only about twenty feet wide by thirty five feet, not very large, but sufficient for our purposes. The section that adjoined the bakery was cornered off and would be suitable for a small kitchen area and a washroom, jointly used with the bakery. It was obvious that before anything else could be done a considerable amount of work was required. Suitable shelving had to be built and the whole area required painting. We needed a show case, deep freeze and pop cooler, a proper till, and meat slicer, to start. We would be required to register our business and arrange with wholesale dealers to provide us with supplies. The rental would be $150 a month, and we would have to construct the shelving units and paint the space ourselves. Someone suggested there was a good chance of finding suitable used store appliances and equipment in Edmonton from a dealer who handled a large number of these types of items through store closures and bankruptcies. I did contact such a company and they did carry a large variety items that would be suitable for our business. We also contacted MacDonald's Consolidated Wholesales who would provide us with groceries. I learned that salesmen came to town on a weekly basis taking orders for meat, pop, fresh produce and confectionary items.

While all this was going on, we were really enjoying our new freedom of transportation. Having a brand spanking new car gave us the opportunity to really get around so we sure took advantage of the situation. On my days off we would frequently take off to visit the folks and family in Vilna as well as camping.

Annabelle and I went into Edmonton to shop around for equipment and supplies. On these trips we would leave Tony and Pam to stay with our parents for the day. We knew that starting the store would cost in the neighbourhood of two thousand dollars. This was an amount we were able to handle, using some of the money we had put away from the sale of our chicken farm operation. We were able to locate a large distributor who handled new and used store equipment and supplies. By the number of items on hand we knew we would be having no difficulty in securing the necessary equipment we required. A considerable number of items were used, but reconditioned and carried warranties. Prices seemed to be reasonable. We said we would be back once we had the building ready. We also made a visit to Macdonald Consolidated, the large grocery and confectionary chain and wholesaler. Getting back to Lac La Biche we contacted Mr. Patry and drew up a two year lease provided stating we paid for our own power needs. Now I had the keys and on my days off started redesigning the building.

Harvest Help at the Farm

While this was happening harvest operations were in full swing at our parents farms. As I had often done I had the desire to help them out during this very busy season. Putting the building we had rented on hold for a couple of weeks, we went to the farm to assist on my two weeks of annual leave. Mom and Dad were always pleased to have a couple of extra hands to help them out. The cutting of the grain and having it stoked had all been done; the threshing is what we would be helping out with. Dad owned and operated his own threshing machine and the tractor that ran it. A few of the neighbouring farmers would help dad and in turn he would thresh their grain for them. Tony and Pam always enjoyed going to the farm and be around the animals. Every morning they would get up early and accompany Mom to milk the cows and they would feed the numerous cats, which lived in and around the barn. My ten year old twin sisters Laurettte and Lucille enjoyed many playful hours with them as well. Annabelle and my sister Marge kept busy helping Mom with preparing all the meals and two lunch breaks daily. Brother Frank who was always a great help around the farm was off to attend U of A in Edmonton, but came home on weekends. Brother Stan who worked in the oilfields at Redwater would come home to help if he had days off. Brother Mike was attending the University of Calgary so it was not very practical for him to attend the harvest operations. With good weather a great deal of the harvesting was completed during the two weeks we helped out. Our payment was also a trunk full of fresh meat and vegetables from the garden.

After returning from the farm, serious steps were taken to have the space we'd leased shaped up in order to start our business. The biggest job was to construct shelving. Once that was finished, we painted the whole area white. I was also working full time so this process wasn't quick. Finally the time did arrive when we felt we could go ahead with purchasing the equipment for our store. On the next trip to Edmonton we acquired a large refrigerator display case, water cooler, pop cooler, cash till and 21 cubic foot upright freezer. These items were then transported by Lac La Biche Trucking. On arrival I arranged them along with a new built in counter area next to the display case. By making a few inquiries I was able to come up with a list of suitable product to start filling the shelves and cases. When we finally did open, it was a very slow start. Many people were unaware of the new confectionary in town. Somebody told us that Coke would likely provide us with a neon sign on our business if a request was made, which we did. The name we registered was "Ann's Confectionary" which was very appropriate. Coca Cola sent us the sign. After it was properly mounted there was a considerable increase of customers, some paying and others just curious.

Both Annabelle and I were very new in this kind of venture. Hours of operation were from 11:00 a.m. to 11:00 p.m. seven days a week. While I was at work Annabelle would be running the store and Tony and Pam would be with her. We had a small kitchen area in the back which was helpful. During periods I was off shift or on days off I would take over these duties. This would give Annabelle some time with taking care of many of household chores that of necessity were neglected. As time went on we both became more comfortable in managing this business. Many different salesman dropped by to take orders on a variety of supplies including everything from cookies to garlic sausage which we sold in abundance. Central meats came in on a weekly basis and provided us with the garlic sausage. In the months that followed we noticed a considerable increase in our business. By no means did we earn a fortune and when we considered the long hours, our net income was not that substantial. It did however, have some perks. Our grocery needs were being provided without a cost factor. Phillip Patry was pleased with us having the confectionary - he noted an increase in his baking sales from increased traffic to the building and we also handled his bakery items in our store. There was always the smell of fresh baked bread in Ann's Confectionary.

Embarras Portage: Before, and After

Being in business also gave us the opportunity to meet many people, and with some we became friends. One couple we became very good friends with were Cam and Sharon Evans. Cam was a social worker with the Department of Welfare. They had four children similar ages to ours. Either Cam or Sharon would come to the store in the evening and spend a lot of time visiting. Our friendship grew and we would visit each others homes frequently.

Lac La Biche Aeradio

Working at the airport was pretty routine, and usually nothing happened that was that exciting. The day shift was the busiest. Watching the sky and preparing hourly weather reports and broadcasting them twice an hour were the main duties. All aircraft movements were also closely monitored and reports submitted to the air traffic control centre in Edmonton also to the next reporting station. During the day shift air traffic in many cases was at times quite substantial but dropped off considerably during the late evening and midnight shifts. One event that stands out in my mind was getting a flight plane out of Fort McMurray. The aircraft was a small Piper Cub with pilot and passenger on board. Most small aircraft, prior to the road being built, flew via the NAR railway. The flight times were two hours and thirty minutes with three hours of fuel on board. Half an hour out radio contact was made with the pilot. He was flying at two thousand feet and was encountering some turbulence. I radioed back providing him with the current weather conditions but also indicated we had a strong cross wind at the airport currently present. He acknowledged and would contact me again prior to landing. Five minutes from the airport he again contacted me. At this point I noticed the wind was gusting up to thirty five miles per hour and a direct cross wind. I reported this and advised him to land at his own discretion. As he made a low pass over this airport, the pilot had encountered some difficulty in keeping this small air plane stable. At this point he again completed a circuit and advised me that he would land on the taxi strip against the wind. As I watched this manoeuvring I could only hope he completed a safe landing. I could see the pilot line up his small aircraft with the taxi strip but there appeared to be some drift. The plane was pushed off course, landing off the taxi strip in a grassy area. The aircraft was still moving rapidly when it hit a small drainage ditch. The wheels were completely sheared off, but there was no other visible damage to the airplane. The airport manager was promptly advised of this incident as well as air traffic central. The pilot and passenger came into the station and did not appear to be overly upset. No one would ever know but I was told that a near empty twenty six bottle of rye was found in the aircraft.

During the fire season our airport was busier than usual. There were two large water bombers stationed. Also in the event of fire in the area there would be a constant movement of crews to the various fire locations.

Some of the Lac La Biche Aeradio Staff

On our staff there were two single operators. Roy Worthington had served at stations in the north and was given a break from isolation. Roy was for a time one of my operators in Fort McMurray. Prior to this he was stationed in Wrigley and isolated station the Northwest Territories on the McKenzie River. A funny story about him comes to mind- Roy had bought himself a Cessna 140 airplane and did a considerable amount of flying in the north. He was going pretty steady with one of the local gals so was in need of a car more than an aircraft. Roy was on leave, flying to see his family in B.C. but ran into some bad weather and could not get through the mountains. He made a quick decision to sell the airplane, buy a car and carry on with the trip.

There was also Jeff Jacobson who as well had some service in the north under his belt. He had been in Lac La Biche for over a year but submitted his application for more northern service. Jeff was hoping for a placement in the Yukon with his preference being Whitehorse or Dawson City. Like Roy he was also courting a local school teacher. His plan was to get married and then transfer to northern station where married quarters were available. Jeff was a hunter, but also possessed a certificate in gunsmithing which he did for a hobby. In our department there was always a great deal of transfers from one location to another. This often involved single staff, mostly to do with northern locations. The action did not exclude married personal but in most cases transfers were a result of promotions. One would constantly run into operators on the move. A good example was again meeting up with Lloyd and Rose Clark. I'd originally met them while being stationed in Embarras. Lloyd and his family were sharing married quarters in the second duplex from ours. They had been in Lac La Biche for nearly two years after being transferred from

Red Deer. A senior operator position opened up in Fort Nelson B.C. He was successful in securing this and they transferred out only a few months of our arrival in town. This was disappointing as we shared a part of our history together. Tony and Pam would sneak over to see Rose who always offered them a variety of treats. Next door to our duplex in the single unit, Arnie Ambie and his family resided. They had moved from Fort McMurray as well where he was airport manager. Similarly to what I had done, Arnie chose to move to Lac La Biche to an airport where the pace was a lot slower. While in Fort McMurray Arnie married a widow with five children. They then had two children of their own. The Ambies were long time residents of McMurray going back to when Annabelle babysat their children. They were now teenagers with the eldest daughter planning marriage. We struggled though another long and cold winter while I kept busy at work and covering off at the store. Poor Annabelle also had a very hectic pace to contend with between the kids, her duties at home and most of the time also running the store.

Town of Lac La Biche

We enjoyed the lifestyle in Lac La Biche even though it did gain a reputation as one of the most violent places in the province. This was not noticeable to us. I guess we pretty well kept out of the night life by having our circle of friends. It was perhaps true in many respects due to sort of a powder keg mentality, being an area with a variety of cultural differences. About one third of the population were Métis or first nations, a third Lebanese, and the last group a mixture of various ethnic groups. Quite a number of the Lebanese families moved into the area in the early thirties to start up mink farming. The reason being there was an abundance of fish in the lake which was the main feed used by these small animals. The prices were exceptionally high at the time and theses farmers, like their stocks, prospered and multiplied. On the northeast side of town near the lake was a large portion of unusable land was an area known as moccasin flats. There were a couple dozen of shacks located throughout this area mostly occupied by natives. It was also known to be a hotbed of social problems. The two town cops who policed the area encountered a great number of problems with the residents in the area. The RCMP did not have any jurisdiction in the town property and were only required to serve the rural area.

Unusual Event

Perhaps the most bizarre incident during my time in Lac La Biche was the mysterious drowning of five local residents. As near as it was ever established, it happened toward the end of February one year. One of the local fish plants was busy cutting blocks of ice to store fish and for the shipment of fresh fish. The ice cutting operation traditionally took place about a thousand yards off shore. One day two young fellows and a couple of girls were driving around in an older vehicle. They were from the Lac La Biche area, but moved about between there and Edmonton. One of the guys had his seventy two year old grandfather living in town. These four young people had picked up the grandfather for reasons unknown. With darkness closing in this car load did something- it was never discovered what- that attracted the town policeman to pursue. The car took off at a high speed along the well traveled road which led to the lake. As they entered the lake there was a "y", with one road going around the lake and the other to the ice block cutting area. The police officer reported losing sight of the vehicle before entering the lake, giving up the chase. Upon entering the lake "y" the driver made a wrong turn and instead of going around the lake, he took the road leading directly into cut out portions of the ice. There was a barrier set up with a warning about the danger of entering the area- but at high speed the car crashed into the barrier and plunged into the cut out ice hole.

Other than the displaced barrier, which the ice cutter thought to be a prank when it was discovered the next morning, there was no other sign of the car entering the water hole. The four young people were basically transients so at first, they were not missed by anyone. The suspicion that something grave had occurred grew with the disappearance of the elder. I can't recall his name but he had lived in the Lac La Biche area all his life, and when he did not pick up his pension cheque the question arose of his whereabouts. After numerous enquiries without any concrete results, the RCMP was called into to investigate. There were no leads, so as a last resort one of the RCMP fellows decided to check out the hole where the ice block cutting occurred. The officer lowered a logging chain deep down into the water hole. Sure enough, he dropped it on top of the sunken car in its watery grave, the only evidence a clinking noise from the chain that vibrated in his hands. The death of five people was not the ending investigators had hoped for, but the mystery was solved.

This finding brought many curious people to the site, but the case wasn't over yet. The final problem was to have this vehicle lifted from more than seventeen feet of water. The early March weather was warm the next year, which made the operation easier. A large crane was brought in and positioned to lift the car from the lakes bottom. It seemed the whole town was there to view the tragic event. The crowd was kept at a distance in a cornered off area near the lakes edge. The crane operator worked with local fire department members and RCMP. At first there was some difficulty in securing hooks for the vehicle. Eventually the hooks caught and slowly the vehicle was brought to the surface. The RCMP quickly covered the entire vehicle to shield the interiors gruesome contents, and it was quickly towed to the local funeral home. In the weeks that followed an inquiry was held to try and established the cause of the tragedy. As part of proceedings it was necessary to determine a possible date for the accident. Our department kept hourly records of weather conditions and as senior weather observer I was requested to testify. I was able to outline to the inquiry boards approximate weather conditions at the time of the accident; weather conditions were normal and no obstructions to visibility would have occurred. The concern was that if there was some obstruction such as fog or snow, the driver could have mistakenly taken the wrong turn in the road. The purpose of the inquiry was not to place blame on anyone in particular but to perhaps make recommendations in the hopes of preventing similar accidents. One of the final recommendations in this sad case was related to ice cutting operations, that the cut out areas were to be properly secured so that the entering of unrelated vehicles was not possible. The funerals for the five unfortunate souls were held soon after the bodies were recovered. It was never discovered why the vehicle took that wrong turn into the icy lake.

Our B.C. Family Visit

Anne and I hadn't taken a break in some time and were longing for a holiday, preferably to B.C. for a visit with Ann's parents. The dilemma we had to contend with was what to do with the store. Ann's close friend Gerta had filled in occasionally but three weeks or so was not an option. We knew Frank was out of the University of Alberta for the summer and helping out on the farm. I hoped Frank would be interested in earning some extra cash to go back to university in the fall. I contacted him and after talking it over with Dad he agreed to help out for the month of July. We were excited to hear this and quickly made plans to travel to the coast by train, which was a very scenic and relaxing way to travel, still is. Frank arrived in his old car a few days before our scheduled departure. Annabelle and I gave him a thorough briefing on how the store ran. He liked the idea of running our confectionary for the summer, so we felt comfortable leaving him in charge. Frank would be staying in our home and of course, his groceries were provided free of charge. We departed by car to Edmonton, left it at my aunts place and excitedly boarded a CN train for the coast. We traveled by train on a number of times before. Our children loved the sleeping births, one each sharing the bed with us. The gentle sway of the train and the meals on white table clothes is something they remember clearly. It took a few days to reach our destination, where we boarded a bus for Powell River. The two weeks at the coast was filled with many hours at the beach and just lying around in the sun. There was a number of successful salmon fishing trips and plenty of oyster picking. Oscar, Mary and Helen enjoyed our visit especially with the kids. Time flew as it does on vacation and we were soon homeward bound. We were wondering what surprises were waiting for us on our return. Frank had done very well considering the lack of experience he had in the confectionary

Embarras Portage: Before, and After

business. To make things more interesting he had even mounted a large speaker outside the store which blasted loud music heard by anybody near the store which was fine. The most noticeable mistake was sort of a humorous one- fifty cases of apples that were piled up in the stores limited storage area. A salesman had smooth talked the inexperienced Frank into ordering this amount by giving him a very "special" deal. We later learned it was the same deal he had offered to all the other merchants in town, and been turned down. We lost most of the apples but were pleased to have Frankie around the rest of the summer helping out. It relieved some of the pressure on me and especially on Annabelle.

The town fathers of Lac La Biche organized and planned the first annual powwow to be held during the August long weekend. A powwow was a ground display with the inaugural theme of the early lives of the native people in the area. Ann's Confectionary provided a float that year, we received second prize for our effort.

Meanwhile our friendship with the Evans family had grown to the extent we would see each other a number of times per week. Either Cam or Sharon would come to the store and spend the often long afternoons chatting with Ann and the kids. If I worked the evening shift at the airport Cam would drop by for a visit. Even though he found my job interesting, he was not that impressed with the shift work or being confined to an office space. He frequently spoke of his work as a social worker, which he loved. Coming from an RCMP background there was a number of similarities on the field handling problems. From what he shared, there was a great deal of freedom and a wide variety of people he came into contact with. Cam dealt with every possible social problem that could be imagined, and helping people to lesson or overcome some of these problems gave him a high degree of pride and satisfaction in his work. Social workers were required to travel sometimes considerable distances to visit with families that were encountering problems. Specific areas were assigned to a number of workers in the office headed by a manager. Cam was one of the four social workers in the office. Each handled a variety of client caseloads. Once assigned a certain frequency of contact had to be maintained with the different clients. Cam did not possess a social worker degree, but was hired based mainly on his experience as a former RCMP officer. After a number of chats, I became intrigued with the idea of social work. Sensing my interest, Cam suggested that I had certain qualities that could qualify me as a social worker. Some training would be required but this was all possible. I did not rush to any conclusions, but in the back of my mind I was giving it some serious thought.

Life and business went on as usual. Next door to our confectionary was a log building. The smaller area of this building next to us was rented by a fur buyer named Sam Woole. This small man of Jewish decent was a very shrewd operator dealing mostly with native trappers. For one thing he possessed a large windfall of trinkets- watches, knives, jewelry, and similar items which many of the natives found attractive. He would trade these items for fur catches brought in for sale, no doubt at a very good profit. Every so often he would bundle these furs and have them shipped to New York for auction. Sam was from New York himself and every summer during off season in fur buying he would travel there for a visit.

The portion of the building next to Sam's housed a second hand appliance sales and repair operator. Mike Makowich frequently came to our store. He told me he resided on an acreage bordering the west end of town. This property was 20 acres in size, but about six acres of this could be easily divided because of a proposed road crossing it. This made gears to start up in my head. I was always looking for a challenge and even though we were very busy between work and operating the store this was something else to consider. I asked Mike what price he would be asking for this six acreage and he claimed it was one thousand dollars. After looking over this property both Annabelle and I felt it looked like a good deal. Mike agreed to be paid fifty dollars a month and we would not get title to this until full payment was concluded, which it soon was. The pie shaped piece of land was mostly covered with large poplar trees with a few clearings scattered around. Two roads bordered this acreage with a proposed road crossing on one end of it. I made plans to have it partially cleared and install power with a view of putting some sort of a structure. In Fort McMurray I still had my prefab ex-chicken barn piled up on my property there. I started thinking it would be a good idea to have it moved to Lac La Biche. We were into the second year of our lease at the confectionary and there seemed to be indications that Mr. Patry would not be giving us an extension. We were doing quite well in the store and it was rumoured that he was considering carrying on with the business himself. It was disappointing, but in a way we felt comfortable letting go of the business. It was hard on Annabelle trying to look after Tony and Pam, running a home and being tied down long hours at the store. Despite spending so much time in the store Tony and Pam were basically a couple of happy kids, having a number of playmates and we'd set up a very secure playhouse area for them in the rear. As suspected, Mr. Patry did come to the store about a month and half prior to the expiry of our lease and informed us he would not be granting an extension. Yes, he would be continuing to run a confectionary under a different name and no, he did not wish to purchase any of our equipment. The refusal to purchase our equipment was a nasty sort of action, and meant we would be stuck trying to dispose of all equipment and excess supplies ourselves.

Confectionary Closure

It didn't take long for word to get around that we would be closing Ann's Confectionary. Many of the people in town who knew Phil Patry had few kind words to say about him. Others who worked at his bakery thought even less of the man. We couldn't disagree. I made inquiries about selling some of the equipment which included a refrigerator, display case, the pop cooler, cash register and meat slicer. We discontinued ordering anymore supplies. We would hold on to our upright freezer and the groceries to take home. Annabelle made a deal with one of the local merchants to take a lot of the surplus grocery items off our hands at the wholesale prices, which we would be able to apply on grocery items we would be purchasing in his store. We were also told about a fellow starting up a country store in Caslan area with the likelihood he may be interested in purchasing some of our equipment. He certainly was interested and a few days later came to see for himself that we had for sale. After carefully looking everything over, we were surprised to hear that he wanted to purchase all the items for sale. The prices we suggested were reasonable to him and very close to the prices we had originally paid. He told us he would have them picked up a day after we officially closed the store. The selling of this specialized equipment, and at a good price to boot, was a huge weight off our shoulders. When the date of the expiry of our lease approached we were ready for closure. A good supply of the dry goods had been moved to our basement. The final items to be moved were the appliances which were picked up by the country store owner from Caslan.

 We were looking forward to finally having some free time to enjoy as a family. After we locked the doors of Ann's Confectionary for the last time, Mr. Patry did take steps to keep the confectionary alive. He brought in equipment and supplies of his own. The Coca

Cola people had the Ann's confectionary sign removed and the bakery had one of their own installed. It was well know that Phil's bakery did have a good business because he did have very good bread items and being fresh daily his products moved well. His people skills left something to be desired thought, and knowing the importance of personal relations in business I did not expect the outcome of his new venture would be good. Although we did not wish this man any bad luck, knowing about his angry disposition it did not appear too promising. Our predication was accurate; in less than six months Phil Patry decided to discontinue his dabbling in the confectionary business. Low turnover combined with poor sales was the official reason, so the space was reverted back to storage.

Tony and Pam were somewhat disappointed about losing the store. They'd always had easy excess to the candy and ice cream which they liked to share with their little friends who hung around the store. Tony was getting older now attending school so there was no lack of playmates to hang out with. With no more worries of spending twelve hour days at the confectionary we began to have some extra time and comfort in our lives. After working seven days a week and a review of our records, it showed we'd only had a four day closure in two years of operation.

With what seemed like a great deal of extra time on my hands, I often took Tony and Pam wandering around on the six acre parcel of land I'd purchased. I was exceptionally pleased with this peaceful chunk of property. There was a small pond next to the road. We cleared off a suitable spot for camp fires with a view of wiener roasts in our minds. At home I drew up a rough plan for a building site with the most suitable level area being in the southwest corner. It required some clearing first. A fellow, Felix Rosslie owned a D6 cat that did the light type of work that I would be requiring. He came and looked over the site and agreed to take on the job. I felt about an acre would give me the necessary space to start off with. He charged $35 an hour and it would probably take about six hours to complete this task. All the trees and other debris were moved to the southern part of the property leaving a good area of nice clear level land. Other than moving my prefab building from Fort McMurray I did not have any further plans in mind for the space. One thing I would have to do before anything would be get power installed on the property. There was a family living nearby so the only requirement would only be one power pole which according to Canada utilities would be a small job, costing about two hundred bucks. Instead of building a new structure I felt the most sensible plan was try and locate something that could be moved. I had a lead on a small vacant building located very near to our duplex. It was supposed to be used as a workshop and was originally a garage. The owners had developed some health problems and so the building has been sitting unused for the last year or so. I walked over to the owner's residence and he told me it was not used and yes he was prepared to sell it. The building was 18 x 20, well built with a shiplap exterior finish. The interior was fully insulated and covered with a paneling on the walls and ceiling. The shingles were still in nice shape and the floor made of plywood. There was a single door and three small windows and even had basic electrical wiring. I asked the owner what price he was asking. Three hundred dollars was the reply, to which I agreed. The fellow who cleared my land, Felix Rosslie, was known to be a local jack of all trades so I contacted him about moving my newly acquired building to the acreage. After a full inspection he suggested the best way to go would be jacking up the building and mounting it on skids which could be dragged. Full length skids could be found at the Canadian utility yard, actually they were used power poles. With a helper, the building was jacked up and the poles secured. Being the early part of winter we already had a sufficient amount of snow on the ground. Skidding along a roadway was a bit of a concern to me but Mr. Rosslie did not think this would be a problem, being the road was gravel base, well frozen and a had a good cover of snow over it, he didn't anticipate any road damage. The move would also occur on a side road on which there was minimal traffic. One Sunday afternoon when I was not at work, Mr. Rosslie suggested it might be a good time to undertake the move. I asked Cam and a couple of other fellows to be available to provide with any traffic problems at a couple of intersections. The building was hooked up to the D6 Cat and to our relief glided along the road easily. In about an hour it reached its final destination, and was placed on a cleared out area we had prepared. A job well done and the total cost for the move about three hundred dollars. So far my total investment was six hundred dollars on this building, an excellent deal.

From our friends in Fort McMurray, we kept hearing about the oil sands development and the way the town was growing. An all weather road was being built with contractors working from both ends. This would provide a major upgrade in the quality of life for people living in Fort McMurray- in fact, if Annabelle and I had known this would happen we probably would have stayed in the area. The original road between Wandering River and Fort McMurray ran a distance of some 250 km. and was rough but passable. Now the base was nearly complete with the exception of 20 km of bush road around the middle. The Wellmans asked us to visit them for Christmas. I could see how excited Ann was on the phone when Edna brought it up. Well, we thought it was a great idea, and it would be exciting for both of us to see all the good friends we'd left behind. I still had my lots there and my prefab chicken barn on the property which would be good to have a look see.

Christmas In Fort McMurray

November and December in Lac La Biche was a cold one usually ranging between thirty and forty degrees F. Our visit to McMurray was on. I had booked a few days off and by working out shifts with a couple of the fellows was able to spend a full week with the Wellman's. It meant I would have to be back to cover some extra shifts during New Years. Annabelle did not like to idea of traveling on the newly developed road with the children especially in the extreme cold weather. She suggested that she and the kids would travel by NAR passenger train and I would go by car. I would have a good idea on road conditions and if it was OK we could travel back together. Annabelle, Tony and Pam left by train and I departed early the next day. A young fellow I knew from the store had relatives at the Fort so he asked me if he could come along on the trip. I was glad to have the company for this maiden journey. Our car although still fairly new I had checked before I left, not wanting any surprises of a mechanical nature along this rugged winter road. Everything looked fine. The tires were almost new and I made sure to pack a good survival kit including plenty of sandwiches and thermos of hot coffee. With the temperature still in the 30 degree below range but with little wind and clear sky's we were off at day break. The well traveled road as far as Wandering River was in excellent condition despite a considerable amount of snow on the ground and along the sides of the highway. It was twenty miles beyond Wandering River where the new road construction began. We drove in the silence of undeveloped bush and muskeg country with few signs of life. The road was narrow and had many ups and downs but the frozen base was fairly smooth with top speed of about forty miles per hour. There was the odd vehicle coming from the north and often there were two or more traveling close together, likely a safety factor. Finally we reached the end of construction after that was where the bush road began. We stopped for a lunch break than proceeded very carefully at a speed of no more than ten miles an hour. It was narrow, rough with constant twists and turns through heavy bush and an occasional high hill to contend with. It took a good hour to once again enter a road that appeared civilized.

About 6 hours after leaving Lac La Biche we pulled into this northern town. I was amazed at the transformation, and looked around interestedly, soaking in all the changes. After being away for over two years and because of the oil boom it hardly seemed like the same place. The Wellman's were no longer living in the Government Duplex because Pete was now working for Suncor in the parts department. After dropping my partner off in town it did not take long to find the Wellman's new residence being it was only about a block away from the duplexes. Annabelle and the kids were happy to see me arrive safely and it was a pleasure to see Edna, Peter and Edna's dad George. George was my trapping partner from Embarras but had left the trap line quite a few years earlier because of poor health and was living with them. The Wellman's adopted children Cathy and Kendall were a few years younger than Tony and Pam but they seemed to adjust well to each other. Since we were arriving in town only two days before Christmas, they kept us busy with festive holiday activities and simply visiting with our old friends. A trip to the airport was also on the agenda, and there some noticeable changes had occurred. I also made a check of my lots and prefab building which to my relief were still intact. Christmas day was unique. All the four children of course were busy enjoying the numerous presents they received. There was a houseful for Christmas dinner that night; our crew, a few of Edna's brothers and their families. The merry meal went on well into the evening. The weather remained brutally cold and we were hoping things would improve by the time we were ready to head for home. This was not the case, so two days before the New Years Eve with the temperature still nearly 30 below F. we made plans to depart at first light. Despite the rough condition especially of the bush road on my trip up, I told Ann we would be able to make the trip home all of us together in our car.

Nicholas Verbisky

We felt well equipped for our frozen journey back to Lac La Biche. Everyone had good winter gear and a full trunk of survival items in the event of a breakdown. Edna made sure we were well supplied with sandwiches and hot drinks. The first hour or so our trip went without a hitch. Little by little we began to notice the windows in the car fogging up and maintaining proper heat in the car began to be a problem. I had to stop a number of times to scrape off the windshield just to have good sight of the road. By the time we entered the undeveloped road through the bush we were encountering pretty stressful circumstances. It was very slow moving through this unforgiving part of the road and that did not help any. We felt totally isolated from the whole world and in limbo. To ease some of the stress we pulled off the road into a safe area and enjoyed Edna's lunch. By the car not moving and the motor running we had some clearing of the windshield occurring which helped. We passed only two vehicles on the road until that point. Upon leaving the bush area we once again got on reasonably good roads, but still not travelling more than 40 miles per hour. Ann and I felt an overwhelming relief to finally arrive at the Wandering River service station. The manager seemed very surprised to hear we had driven in from Fort McMurray in this kind of cold weather. We all were able to enjoy hot chocolate and a few snacks. The car was topped off with gas and we departed in much better spirits now traveling in a more civilized part of the country. Arriving home a total of seven hours later was a joy.

I worked the afternoon shift on New Year's Eve so there was little celebration around our house. On New Year's Day we had the Evan's family for a visit and an after dinner meal. As usual work was at standstill due to an almost complete shutdown in air traffic. I still had to complete hourly weather reports for broadcasting and dissemination which was about the only task required.

My friend Cam dropped by one particularly slow afternoon and again tried to persuade to consider entering a career in social work. He'd even brought a number of pamphlets with training information about this career. As a great deal of my work involved sitting and waiting, over the rest of the winter I had a lot of spare time on my hands. It was during this time that I started to review the various information that Cam brought in for me to read. The more that I read, the greater my interest became. With social work there was every imaginable human problem to contend with. It brought back memories of when I finished high school and my only goal in life was to join the RCMP. It was the most disappointing experience when I failed the personal interview after passing all the other exams. To this day, I still blame the British personal officer who did not take kindly to applicants who were not of strictly English origin. The past is the past, but when I went over the various duties of a social worker, there seemed to similarities to that of an RCMP officer, and that piqued my interest further.

Lac La Biche Trucking was the service we used at our confectionary in bringing in our supplies from Edmonton. Talking to one of the Wowk brothers who operated the business I learned that they had been making occasional trips to Fort McMurray with cargo equipment and supplies on one of their flat bed trucks. He mentioned there was a truck that often returned empty. Another plan came to mind. I explained to Gerald that I had a dismantled prefab building at Fort McMurray that I wished brought down. He felt it could be arranged and the price of delivery would depend on the amount of work required for loading. I suggested I could have someone available at that end to assist. Raymond Cardinal, the Metis man who lived next door to the old chicken farm had earlier asked about the storage shed on the property which I gave to him free of charge. In return he kept on eye on the property and would likely help out with loading. Peter Wellman is another fellow who would help if need be. I gave Gerald both these names as well as the location of my lots. Gerald said once the weather warmed up the new road to the north would be shut down. The twenty miles of bush road would become unserviceable due to a lot of low land and muskeg. It was February when I received word from Lac la Biche Trucking that my cargo had arrived safely. Cam and Ivan, one of the radio fellows, came to assist. Delivery was made to the acreage and unloaded easily. I was very pleased with service Gerald had provided at a total price of three hundred dollars bringing my total investment on buildings at the acreage to about $900. Now that I had this building as well as the other one previously moved to the acreage, the question was what would my plans for be them? I knew that the town had encountered problems with not having enough rental accommodation available. This gave me an idea of having both these units upgraded to a proper standard and then renting them out. This would involve a great deal of work and considerable cash infusion but could provide a nice return on the investment. We had a fairly nice savings built up after closing down the store, but now funds were running low, so borrowing money from the bank would be required to do the renovations. The prefab building structurally was in excellent condition, but because of its previous use as a chicken coop obviously total disinfection before any work was considered. With the plan for my property in place I would now be waiting till the arrival of spring and warmer weather to become more actively involved.

A Social Worker Experience

Cam asked me join him on one of his trips to some of the isolated outposts he visited. A considerable amount of the area he covered was along the northern Alberta railway route extending halfway to Fort McMurray. One day I agreed to join him on one of these excursions. It was a three day trip, planned for mid March. We travelled to the small community of Chard by train. Train travel was the most effective way to go. The forestry road to the area was in poor condition and in many cases impassable. We left on a Monday afternoon and packed sleeping bags. No hotels for social workers- we would have the use of a forestry cabin in Chard. The life of a rural social worker did not seem particularly glamorous. The five hour train ride *was* very interesting despite the desolate and remote landscape.

Lac La Biche appeared to be a major boarding point for many of the families in the area. People would arrive from the various points along the track doing their shopping and other business then catch a ride home on the returning trip. The train itself had a character all its own. There were a number of older passenger cars, a baggage car and a couple of freighters. As we rode along at a fairly slow pace there seemed to be a considerable amount of entertainment taking place within her cars. Beer was consumed openly and at one corner a poker game was in progress. A friendly porter walked around, smiling and shouting, selling pop and confectionary items. In our seats Cam opened up his brief case which carried a number of files and he started showing me the case logs for some of the clients he was planning to visit. There were a number of issues these individuals and families were dealing with, ranging from minor to more serious. He had a voucher book which would be issued to families that required financial help, single mothers and others for people who were ill and unable to secure any type of regular work. Some people had medical problems and required assistance with transportation. Child welfare concerns were a major problem in many of the native families in our area. There had been cases where Cam had to apprehend children and try to place them in foster homes. Many of these problems arose because of alcohol abuse. Apprehension is usually a last resort and even then, often children are placed with relatives such as grandparents until the parents sober up and able to again take care of their children.

I had a chance to take in what there was of scenery out the large glass windows of the train. Every so often we would have a whistle stop coming across a small village or haphazard section of residences occasionally built up near the tracks. A fellow by the name of Mike Stepanowich ran a trading post near the settlement of Chard. Cam had earlier advised Mike of our coming out to the area. People that were not on his list but required government help of one sort or another also were made aware of Cam's coming arrival and would plan on meeting him there.

Chard was about ten miles from the railway section stopover which was the end of the line for us via train. We were met at the Chard drop off site by Mike who also met the train to pick up supplies for his store. Jim, a forestry officer was also there and he too had items to pick up during this short stop over. The forestry cabin was near the Stepanowich store where keys to the cabin were left for us. We loaded our gear on Mike's truck and after he loaded his supplies we were off down basically a trail to his trading post some ten miles away. His store was on a hill near the small Indian reserve. There were treaty Indians living on the reserve, but in the same general vicinity also near the lake lived non treaty residents as well. Cam had no jurisdiction on the reserve other than for child welfare matters. He was however responsible for the Métis (non treaty) families in the area. There were about one hundred or so people on the reserve and almost as many in the non reserve area. The trip to the store took some time- travel was slow due to the rough road conditions. It was late in the afternoon before we arrived at the cabin. It was a rough hewn small, one bedroom affair with no neighbours close by. After grabbing our keys Cam decided we would not be making any home visits until the next day. After dinner, which included some fine steaks we'd brought along with us, we dropped by to Mikes store for a visit and to get an updated list of families that wanted to see Cam. There were no less than twenty families that wanted his attention. Cam would rent Mikes truck to get around in the scattered community. After composing the next days plan, we decided to hit the sack. Despite the primitive lodgings we were glad to finally settle into our sleeping bags after a long day.

Morning arrived early. We started the day right with a hearty breakfast of bacon and eggs cooked on the small propane stove. So far I had found this trip exciting and deep down I believed this was the type of work I would like to do. Sometime after eight we got Mike's truck and we were off to make our first home visit. Before that we made a general sweep of the whole community. Most of the dwellings were log constructions and according to Cam the smaller sized units had two rooms while the larger ones had four. The families last names included Cardinal, Auger, Jacobs, Desjarlais and Bourgues. Cam said that a large number of these families lived in common law relationships, and trading partners was not a big issue in these parts. Our first visit was to a Mary Auger, a single mom with three children. Mary was only thirty five but seemed much older. Cam issued her a grocery voucher to the outpost store as well a note from a health nurse, who also visited the area monthly, indicating that her ten year old boy who was suffering from unusual stomach pains should be looked at more closely so train fare was being provided for her to go to Lac La Biche for a medical review. Mary's home was fairly clean with only very basic furniture items in place but seemed adequate for her needs. Cam would usually survey to determine if there was a need of additional or replacement furniture. A second hand store in Lac La Biche was favoured for this and the pieces would be shipped in by train. After a few more visits I noticed Cameron's way of handling matters was very good. Being an ex RCMP officer it was clear he still had a very professional manner about him and exceptional investigating skills. He was also very considerate and compassionate, but firm in his decision. As I attended all home visits with him, people would question my presence. Cam replied that I was a social worker in training. Home after home Cam would sort out the various problems and in each case would make decisions based on the individual circumstances. He issued numerous vouchers to cover groceries, rental, purchase of furniture and medical transportation. There was also child welfare issues with apprehension, always being a last resort, and thankfully did not occur on this trip. It was surprising to see how many of these families lived. Many of the residents were nice and fairly clean and others in fatal disarray. The majority of the clients seemed friendly and cooperative, but there were also a few who were very negative and hard to reason with. On the whole it was clear that Cameron was liked and approved by almost everybody in the area. It was late afternoon when we made one of our final visits to Johnny Auger's home. He had just returned from the lake carrying a five gallon container full of fish. Johnny had a large family and was given a permit by the fish and game to set a net once a week. These fish were mostly pickerel, a few jacks as well. Johnny offered us a couple of these fish, but it was against rules for Cam to accept gifts. I could not pass up the deal so I accepted one of these nice sized pickerels which we enjoyed for supper.

By that afternoon Cameron had pretty well completed all his assigned work with the exception of some minor details he had to clear up in Mike's trading post. As there was a large number of food vouchers issued to the store there was quite a number of families coming to purchase their supplies when we arrived. Mike had a native clerk helping him out during busy periods such as this. The train going south arrived at the Chard section station at noon the next day. Mike would drive us back to the cabin and pick us up again in the morning to catch our ride home. Before retiring for the night we went over to visit with Jim, the local forest ranger. Jim was married but his wife was away down south visiting with her family. We enjoyed a couple cups of coffee with him and thanked him for the cabin accommodation. The next morning we returned to Mikes. A couple of people not on the list were in need of assistance, so Cam processed their applications while we waited for the train. By eleven o'clock we were on the road again arriving at the railway station shortly before the trains arrival. The atmosphere of the train heading south was less jubilant, the passengers more reserved. The train rolled into Lac La Biche close to suppertime which was an end to a very interesting experience. Cameron was well aware of the most unusual couple of days that opened my eyes to people that were so less fortunate then ourselves and how they can be helped to cope with their daily lives. The arrival of the spring was pleasant and brought my family a lot of joy. The heavy snow disappeared and it seemed that everybody was outdoors again, spring cleaning and getting ready for another growing season. Easter festivities came and went with a family visit to the farm. Work at the airport was routine, peppered with a few staff changes. Jeff Jacobson got married and was transferred to Dawson City in the Yukon. A single fellow by the name of Bill Kawasnycia took his place.

Upgrading the Acreage

Any spare time I had was spent at the acreage looking over the workload that awaited me. I would take along my chainsaw to do some fine tuning of the cleared out area. Annabelle, with Tony and Pam in tow, would often be there for company and a wiener roast. I was anxious to start working on my small building preparing it for a rental. The building was basically in very nice condition, and only required painting, a floor covering and a chimney installation for a wood cook stove and a few more homey touches. I'd purchased a small cupboard for the kitchen area, and because of the limited space a partition wall was constructed dividing the building into two equal rooms, one comprising a kitchen area, the other half a common room and bedroom space. I did have some carpentry skills but there were some work that required a professional touch. Brother Frank was home for the summer from university, so I was planning to have him in the completion of the prefab building. Frank was enrolled in a teaching program but his specialty was the industrial arts, so he possessed excellent carpentry skills and other skills pertaining to construction. The power company did install a pole and now I had power on the property. Even though Frank's services were required on the farm he felt he could probably spend a month working on my building which he felt would be sufficient time to bring it up to make a suitable rental standard. Frank arrived after helping dad complete the spring work. The biggest job was to set up the prefab sections which would require some heavy lifting. For this I lined up a couple of young fellows to help out. The ground was nice and level in the spot we chose. Once we got the two 8x8 cedar cross beams placed, the building fell into place like a jigsaw puzzle. The 4x10 foot roof double sheet insulated panels were the most difficult to install. They were very heavy and required to be lifted on top of the building. The only equipment we had was a couple of good ladders and ropes to joist them up. Frank supervised a lot of the project as I was away working my regular shifts. After three days the building was intact and the physical part was now completed. The next step on the agenda was to have this former chicken coop thoroughly disinfected and steam cleaned which turned out very well. Frank updated the electrical wiring and I had an electrician come check that out as well as mount a panel. There were repairs to windows, doors and complete repainting. Two insulated chimney units were installed, one for a wood cook stove and the other for wood heater. Partition walls were built and building was divided into four large rooms. The exterior was painted as well. Frank was able to complete all the work in the time we had. Both rental units were now ready to take tenants. It was well known that the town always seemed to have a shortage of rental units at the time. Word got out that I had two places recently completed. A fellow by the name of Ernie Cardinal approached me and asked about looking at these buildings as he had recently lost his home to a fire and needed a place to live. The welfare department had put him up temporarily in a motel unit. Ernie had a wife and three small children. Ernie felt the prefab building was a bit too large but he like the smaller unit which he felt had limited space but was much better than the motel room. He also liked the idea of living in a country setting which had a good wood supply for the cook stove and his children were free to play outdoors with few distractions. Shortly after the Cardinal's moved in, a social worker from the welfare department contacted me knowing they were in need of a home for a large family and wished to know if I had such a place. It would be for John Desjarlais, his spouse, his elderly mother and four preteen aged children. The social worker looked over the prefab and felt it would be adequate for this family's needs, and the rent was reasonable. With both these buildings now being rented out I now had a bit of extra income. Half was going to the bank to pay off a loan and the remainder a little extra cash for whatever we wished to spend it on. The tenants were a mixed bag; Ann Cardinal kept her small home spotless. The Desjarlais family lacked considerably in that area.

Nicholas Verbisky

With both Tony and Pam now attending school Annabelle had some extra time on her hands. Ann is a very gifted hairdresser. It comes naturally to her, as it did to her dad who was a good barber, and to a large number of her family members in B.C. who owned and operated hair dressing salons. Her friend Angie Coutney owned a hairdressing shop in town and offered Annabelle to get her certificate through her shop knowing how capable she was in that line of work. Having an empty nest at home, she expressed a desire to have another child, therefore was thrown between the hair dressing training she was offered or having another bundle of joy. Things were changing for me as well. After my trip with Cameron I began to feel a much greater interest in the field of social work. Most of the rural social workers did not have degrees in social work. Many did however have college training in other fields, or general very positive life experiences. The largest number had come from police or RCMP backgrounds. This, however, did not always rate as high on the list as applicants who possessed good family relations skills, were understanding and compassionate to the needs of people who were less fortunate. While at work I had a lot of free time so I studied in great detail the endless supply of material that came available to me. Even though I had a good stable job with good benefits, reasonable pay and a secure position, the idea of being a social worker appealed to me greatly.

As a family we all enjoyed the nice summer months, frequently going on camping and fishing trips mostly in company with Evan's. Our favourite spot was Missawawi Lake, near the Hylo campsite. That was where the perch fishing was excellent and usually followed with a fish fry for everyone to enjoy. It was during this time that we received some disappointing news. Cameron was being transferred to Athabasca. Information was the Athabasca district office was upgrading their staff from two to five social workers, given the large area they were required to serve north of Slave Lake. Cameron was an experienced worker and his services were required. Cam was also very knowledgeable working with native clients and the new assigned area had a large concentration of first nation families. Athabasca was only a three hour drive away, so we knew we would still be seeing them often. It was still a difficult move. In less than a month we said goodbye to our close friends and the Evans family made the move to their new home.

In early fall we made a couple of trips to the old family farm. Harvest operations had begun and a helping hand was always welcomed by the parents. Dad was old fashioned in many ways and still preferred harvesting with a binder and stoking the grain. This was followed by thrashing, a very labour intensive way of doing things. He did not trust the combines as wet weather could be very destructive if the grain was cut down and let to lie in swaths causing it to rot.

Family Increase Confirmed

Meanwhile, Anne had made her decision. Instead of pursuing getting certification as a hair dresser, she felt that the addition of another family member was a great idea. It was soon after that Annabelle did make a visit to our family doctor and it was confirmed that indeed she was expecting. A couple months later she noticed she was putting on more weight than usual. Ann was referred to an obstetrician by the name of Dr. Brown, who was practicing in Edmonton but was known by Ann's family from when he practiced in Fort McMurray. While attending his medical clinic in Edmonton it was discovered that she was carrying twins! As Annabelle was not that large of a person, the extra load was at times getting her down but both of us were delighted about what was in store for our family.

The Evan's still came to visit us frequently. They enjoyed the town of Athabasca, and Cameron also enjoyed working at that district office. We made a point of making a trip to that town as well and from which we seen it was graced by a very picturesque landscape- the rolling hills, the endless fields, the gorgeous sunsets. One of these visits I had the opportunity to meet Cam's district manager. His name was Anthony Pearce. He was an ex-Anglican minister. Cam had spoken to him about me and said that I had expressed a sincere interest in social work. I was later told by Cam that I had made an impression on the manager. It was Mr. Pearce feeling that because of my background, and general life experience, I probably would make a successful entry into the social worker field. He told Cameron that if and when I did make up my mind he would submit my name to the personal office in Edmonton with a positive review.

With the twins arrival date quickly approaching Annabelle was required to make monthly trips to Edmonton to see Dr. Brown. If I did decide to change fields and become a social worker, what would the future have to offer our family? When I looked at the social workers income of $395 per month I had to consider that I earned twice the amount in my job. Our rent on the three bedroom duplex was $92 per month which included all utilities. Cam said rent for a similar unit in Athabasca was about $110 per month and payment on utilities was also required. Our home was fully furnished, included in the rent, but if we moved we would be required to buy almost everything again. My job was a permanent position, very secure with a good medical and pension plan. How would it be to make a move to a much more unknown future? I discussed it very seriously with Annabelle. She assured me that she was comfortable with whatever my final decision would be, and that we would do well in whatever road was taken. I also brought up the issue with my O.I.C. Oscar was very surprised to hear I was considering leaving the Department, and did not think it was the most beneficial step to take, but knowing what type of a person I was agreed I would succeed in any endeavours I undertook. I took some comfort in knowing I could delay the big decision; nothing was final yet as I first had to undergo an interview with my potential future employer which meant I could be disapproved as easily as not.

Decision Made To Pursue Social Worker Career

After a great many sleepless nights and soul searching I finally made up my mind. I felt I would never be happy not knowing what my future held if I did not apply for the social worker position. If I did *not* pass the interview I wouldn't feel too badly and would move on with life. With Annabelle's blessing I contacted Mr. Pearce, the man in charge of the Athabasca welfare office, and told him that I did make up my mind and requested an interview to be held at the central headquarters in Edmonton. In less than two weeks of my phone call to Mr. Pearce I received a letter from the provincial welfare headquarters in Edmonton requesting my presence for an interview. This again created some additional stress which seemed to be the order of the day. The interview was conducted by three personnel individuals, two fellows and a lady. My jitters settled quickly and after a great amount of questions and answers, I felt I had done exceptionally well. From all indications they seemed to be pleased as well. One of the conditions I did outline to them was the time line. If I was successful I would require at least one months notice because of my wife's pregnancy and other issues that required clarification, mainly my aeradio job. They told me I would be advised by letter of their decision. The letter in the mail arrived less than two weeks after the interview. The good news came- I had been successful in my bid to become a social worker. There was a need for a worker in the Athabasca District Office, and my placement would be at that location. They offered me a very flexible period before reporting for duty, fully knowing my circumstances, but suggested it should be no longer than a month.

Embarras Portage: Before, and After

Annabelle was seen by Dr. Brown in Edmonton and after a closer examination it was requested she enter the U of A hospital recommending bed rest with the expected date of delivery of the twins being mid May. It was now past the middle of April 1967. While at work I arranged for Gerta to look after Tony and Pam. She was such a kind and generous lady, and the children loved staying with her. Oscar was disappointed that I was leaving the Department, as I was his right hand person at the station. Talk about a stressful period, this was it. I had to deal with some of the packing for the move and this was much harder without Annabelle's help. Cam said he would look into getting us a place to rent in Athabasca. Cameron's family lived in a three bedroom town house sand there were a couple of additional units still available. He would contact the manager about our move with a proposal date being June 1, 1967. My mind was in turmoil with all changes that were ahead of us. With Annabelle in hospital, caring for Tony and Pam, working, packing and arrangements for the move was all getting me down, but in another way I was happy and looking forward to the new challenge ahead.

Aeradio Position resigned

I submitted my resignation immediately giving the Department one months notice. Some of the headquarters staff whom I had known for years were very surprised to hear of my leaving but wished me the best. With the weeks passing quickly I was surprised the headway I was able to make with packing and making other plans for our departure. A couple of weeks in a row Cam drove down from Athabasca and helped out with the packing. I contacted Lac La Biche trucking about moving us. The operator did not see a problem as he traveled almost daily to Edmonton and could load us and travel via Athabasca to the city. As his trips to Edmonton were usually empty his charge for the move would only be seventy five dollars, a token fee. It helped with Steve being our neighbour and we had given him our business while we ran the confectionary. Annabelle would phone me almost daily about her progress and also, feeling badly about everything at home being on my shoulders. I tried to keep her mind at ease by telling her that everything was fine and how Tony and Pam were doing fine. Annabelle told me how uncomfortable she was feeling, largely due to the excessive weight she was caring and being assigned to bedrest. She was glad to have a lot of support from a few of our friends in Edmonton who came to see her nearly every day.

Our Twins Arrive

On May 17th, 1967 while I was working the afternoon shift, the good news finally arrived. Annabelle was given a phone and with a great deal of excitement informed me we had just become the parents of a healthy boy and girl. The new additions to our family brought us all a great deal of excitement. Early the next morning I was on my way to Edmonton to see Annabelle and our two new treasures. We had talked about which names we would give them but it also depended on whether they were girls, boys or one of each. As we were blessed with one of each we decided on Colleen for the girl and Colin for the boy. Colin weighed 6 lbs 4 oz. and was the bigger one of the two and Colleen being just 5 lbs. Annabelle and Colin would be able to come home in a week but Colleen was required to remain in hospital for an extra week until she put on a little more weight. Traveling back and forth to Edmonton on gravel roads was not the most pleasant chore, but in a case like this it was well worth it. The house was basically in turmoil with all the packing that was going on in preparation for the move which was in less than two weeks. Annabelle still required rest and was adjusting herself to having two new babies. She was very pleased to have a good and faithful friend in Gerda to help her out with the other two. We were happy when we received a phone call from Dr. Brown at the U of A hospital that Colleen was ready for discharge. This was thankfully our final trip to Edmonton, now having Coleen with us for the ride. It was wonderful to have everybody at home even though we would be on the move very soon. Cam had made arrangements in Athabasca to have our belongings unloaded into the town house we had rented which was only a few doors away from theirs.

Our Move to Athabasca

During the last week prior to our move the Kleven's asked that we stay with them. Both the twins seemed to be sleeping a good portion of the time so we were able to complete our final cleaning at the duplex. We were surprised to have a low key farewell party hosted for us at their home. It was attended by many of our friends from work and town wishing us well on our new journey. Our departure date of May 30, 1967 finally arrived. The weather was warm and sunny, after a couple of rainy days. You would never believe the manner in which our 1962 Stratochief car was loaded for the move. Strung across the back was some clothing, and a bird cage hung with a small finch inside. Tony and Pam where huddled into a limited space and on the floor along with Simmi, our family dog. The front seat was cramped with Annabelle, me on the drivers side and our two new additions. The first part of the trip went fairly well but with recent rain the gravel road was muddy and difficult to drive on. Our car did not have air-conditioning and Annabelle refused to have the windows lowered for air as she did not wish the little ones to get a draft. Simmi seemed a bit stressed with the confined area and continued to pass a lot of gas. Between the enclosed heat in the car and Simmi's continuous gas – suffice to say it did not create a very pleasant atmosphere. I had to stop a few times just for us to get a little fresh air.

Once we got close to Athabasca we had to get off the main highway. During the last twenty miles we entered a country road which was narrow and poorly gravelled. Being wet, travel was slow and it was even more difficult to stay on track. We were all extremely pleased to finally arrive in Athabasca two and half hours later. On arrival we were very warmly greeted by the Evans family. I felt a sense of relief inwardly, knowing that we were moving somewhere with our sincere friends to help us through what would be a very stressful time for us as we settled in to our new lives in Athabasca. Prior to this point in our lives, the federal government had supplied us with all the household furniture items we needed. Now we would be required to find our own. We had already ordered some of the kitchen furniture and beds from the Eaton's catalogue. They were shipped to the attention of Cameron, and he had them uncrated and properly placed those in our home. Still, the new space seemed very empty. Thankfully I was given a few days grace before reporting to my new job. I spent the time helping Annabelle to sort everything out, and kind of stabilize everything. The new environment, our new babies as well as Tony and Pam would have her very busy. I knew it would be a stressful time for her settling in to this new life as much as it was for me starting on my new career path. It was also very exciting for both of us, looking back.

Cameron had filled me in of what lay ahead. I would receive an orientation period, sort of an in-service training for a week or so. I would tag along with him acting as my trainer. My new position would require me to serve another portion of the province. One week in home base and another week serving the area of Wabasca, Trout Lake, Sandy Lake, Chipewyan Lake and Peerless Lake. Cam and I would drive the very rugged road to our home base to Wabasca, eight five miles northeast of Slave Lake. There were no roads to the other remote sites so we would be flown in by a Cessna 180 Charter. I pondered over what was ahead in my new chosen career as a social worker. My mind still remained unclear if I had made the right decision. The sense of indecision was closely followed by a feeling of excitement knowing that a new challenge was ahead.

My journey as a Social Worker took me another 25 wonderful years to compete. Consisting of 18 years as a frontline worker with the balance being in a Regional Office setting as a supervisor and consultant; but that is another story.

Our Wedding Day

On our honeymoon

Annabelle and myself, with Tony and Pam

Tony and Pam, holding Colin and Colleen 1967

Annabelle and myself on our 25th anniversary

Oscar and Mary Adami with Joe and Katie Verbisky

Pete and Edna Wellman, Athabasca AB.

Annabelle and myself, with Colin and Colleen

Myself (back row, far right) with Mom and siblings

Oscar, Mary, Annabelle and Helen Adami, Powell River BC.

My parents

Family Portait- top left to right Tony, Nick, Colin. Bottom- Pam, Ann and Colleen

Happiness years later

Immediate family on our 50th at St Paul AB.

Fifty years later with our four children

Epilogue

I am so proud of my family, beginning with my strong-minded and wonderful wife Annabelle. We often generously overlooked each other's faults and kept our initial love alive through half a century. I am lucky to have wife like her.

Our children and their families:

Anthony (Tony) was born November 3, 1957 in Fort Smith NWT. After graduating from high school he attended both the U of A and NAIT graduated as an accountant with the CGA designation. He worked in his field at various locations, but is currently general manager of the Ramada Inn in Loydminster Alberta. He is married to Joan Stratichuk and they have two girls Madison b.d. June 24, 1991 and Rebecca b.d. March 8, 1995.

Pamelia (Pam) was born June 13, 1960 in Fort McMurray Alberta. After high school graduation in Athabasca, she attended Grant MacEwin College and received a diploma in equine studies. Many years later this was followed by nurses training receiving her LPN. Pam married Grant Breckenridge and they have three children, Amelia b.d. July 1, 1983, Lane b.d. December 28, 1984 and Delton b.d. June 8, 1992. They reside in Cowichan Bay B.C. where Pam is nursing and Grant is employed as a building inspector.

Colleen was born on May 17, 1967. After graduation from the Athabasca High School she took secretarial and Rehab Practioner training. She married Orrell Sheets and they have three children, Noel b.d. February 19 1993, Kyler b.d. September 15, 1995 and Hailey b.d. Oct. 1, 1997. Orrell graduated in the field of social work and has worked in various locations in Alberta, B.C. and NWT. He is currently warden of the Adult prison in Yellowknife NWT. Colleen is employed as a rehab practitioner working with the handicapped.

Colin is Colleen's twin. After graduation from the Athabasca High School he enrolled into a two year Law Enforcement program at Grant MacEwan College in Edmonton. Upon graduation he worked as a security officer and later as a social worker before being accepted into the RCMP. Colin was married to Connie Smigleski but they divorced. They remain on a very friendly basis and together jointly are raising their son Colton b.d. January 1, 1994. Colin is currently stationed as a RCMP officer in Invimere B.C.

As the sun sets on my life, all I have are many memories which have so successfully carried me through the years. But at the age of 75 I cannot delay telling of the earlier years, or I will join the long list of those tracks through history that have been swept away by the northern wind and snow. I hope you have enjoyed reading this book as much as I have enjoyed living it.

Both Embarras Portage and Smith River were emergency Airports. They provided a very significant service during their tenure. Increasing technology in communications and the jet age in aircraft eventually meant their services were no longer required. Both bases closed in the 1960's. Apparently the deserted runways still remain as drop off points in event of forest fires in the areas.

As an Aeradio operator, the requirements for this position were very specific. One had to be a holder of a second class license in commercial wireless radio operating. The International Morse Code system was used in this case. This was the major means of communication throughout the northern regions prior to 1960 as technology in communications significantly increase these service slowly faded out. As a result the radio operator's "keys: and "bugs" became silent.

Radio wireless code operation is still used as a hobby by "Ham" operators.

After the completion of my manuscript I pondered who would be the most suitable to review and edit this work. My granddaughter Amelia Breckenridge came to mind. Here is a poem she wrote that was inspired during the process of editing my book.

Winter in Embarras Portage

No dawn breaks on the deep chilled arctic passes;
long midnights bearing down on tundra grasses
landscapes covered with that ancient blackness
of six months night where no sun ever passes.

But inside trappers cabins lamps are shining
on faces lit with company and cheer.
It's not the darkness there that is presiding
but good friends gathered close, kept near.

Amelia Breckenridge